REFINING NATURE THE LANDSCAPE ARCHITECTURE OF PETER WALKER

REFINING NATURE
THE LANDSCAPE ARCHITECTURE
OF PETER WALKER

SCOTT MELBOURNE

BIRKHÄUSER BASEL

CONTENTS

FOREWORD

NIALL KIRKWOOD

This publication demonstrates the richness and breadth of the current land-scape design field as well as introducing and portraying one of its finest design practitioners: Peter Walker.

The drawings and photographs in this monograph illustrate the living dynamic processes of landscape design, with leaps of imagination supported by landscape architect Peter Walker's steady direction within periods of crea-tive design synthesis and integration. As can be seen in the resulting body of built landscape design work that spans across the world, creating and shaping landscape form has required a lifetime investment of craft and invention and requires a clear intellectual focus, a deep understanding of landscape design history with as much practical site experience as any designer and design of-fice can muster in a professional lifetime. In return, a built landscape is made that is clear and strong in its resolution and execution, an art form combining materials, ideas, and meaning, which is simultaneously poetic and pragmatic.

What makes the design work of Peter Walker examined here so original and yet so grounded in place, contemporary culture, and the traditions of the field of landscape architecture is his sensitive combination of natural mater-ials, the immutable, the durable. Design in Peter Walker's landscapes arises carefully from the impact of ideas upon these materials in context. They stem from evolving forms according to surface, function, and texture, the needs of rest and shade, of lawn, of canopy trees, of pathways. They arise from the human wish and need to formulate ideas through design action, to recreate them into entities, so that their meanings will not depart fitfully as they do from the mind, so that thinking and belief and design sensibilities and attitudes may endure as actual things to be engaged with and, most importantly, be enjoyed by all. Peter Walker's influence on the design field and the landscape discipline is unchallenged, the body of built work worldwide is remarkable and, more sig-nificantly, this publication adds to the body of core knowledge about and for landscape architecture now and in the future.

Cambridge, Massachusetts, November 2019

PREFACE

"Dig hole... make mound. This is landscape architecture—it's so simple." [1]

Richard Haag

*"To successfully build a fine work of landscape architecture
is one of the most difficult tasks in the world of design."* [2]

Peter Walker

Both of these master designers are correct. In Haag's declaration, there is the call for an intervention, a reshaping of the ground that marks the beginning for any designed landscape. A seemingly simple topographic operation instantly creates possibilities, there now being a place for water to collect, a change in soils and hydrology expressed through vegetation, a slope to recline upon, a place to gather. Haag's shorthand summary for how to begin the process of landscape formation relates to his ideals of "non-striving," [3] where the efforts of the designer are intentionally understated rather than emphasized through the built work. Dig hole, build mound, don't overthink it—this can be simple.

Simple, but not easy. The divergence from clarity of concept to challenge of implementation could be described by any experienced designer, [4] but when expressed by someone as accomplished as Peter Walker we ought to take notice. Here is a person who has actively been designing landscapes for more than half a century and produced more than a hundred built works, in the process winning scores of awards from those within and also outside of his own discipline. And like star athletes reflecting on past accomplishments, he is reminding us that it is not easy, none of it was easy.

The challenges with any built landscape are diverse and compounding: conceiving and communicating ideas for a medium that is resistant to objectification; the necessary buy-in and financial support from decision makers for something that even if valued might not be considered essential;[5] the sourcing of materials and attempted guidance of a construction process typically driven by a third party; once successfully built, weathering effects brought by environmental exposure and continued use over time; and ultimately, existentially, the shifting needs and values of a surrounding community that may in time no longer consider the landscape worthwhile to maintain or even keep in place.

Built landscapes are difficult to describe, expensive, contested, and vulnerable. To go from design notion to built work to valued community asset—it is one of the most difficult things.

If building a landscape is fraught with so many challenges, what are design strategies that might improve chances of success, however that success might be considered?

GOAL OF THIS BOOK

Driving this study is an essential question: what might one learn from a careful examination of Peter Walker's built works, developed across a range of conditions and geographies over the past six decades? With a focus on identifying design strategies embedded within these projects, the book aims to equip landscape architects and other environmental designers with new insight for effectively shaping inhabited outdoor environments—for building landscapes that resonate and endure.

Each of the discussed projects was experienced and documented in person for the purposes of this study, totaling more than forty sites in six countries on four continents. Each of these landscapes is a multi-layered environment that may be accessed, inhabited, examined, and learned from. This is one of the great advantages for students of landscape—that our source material is *out there*—and my observations from these site visits mark the starting point for the identified strategies. The fact that most of these projects have been in place for many years makes them all the more valuable as they each have stories to tell regarding physical weathering and evolving usage over time.

While these projects were sought out and experienced individually, I have prepared the following reflections and integrated desktop studies with the intent that they be as accessible to others as possible. Design—let alone a lifelong career as a creative professional—is challenging enough in its own right.

And who might you be, dear reader? Many of you will be students of design, no matter what your age. For those still early in your career this text will be especially valuable in offering a distinct set of case study sites while, even more importantly, illustrating how the different systems of landscape may be considered and interrelated at the site scale. My own students have been in mind while creating this book and I have attempted to make the text engaging for readers from a range of contexts. The highlighted design lessons are not dogma, but rather notions that may be embedded into your thinking, ready to be revisited and tested over time.

For more experienced designers there still are discoveries to be made, in addition to new ways of reflecting on your own work. You working professionals will likely have been exposed to some of these projects during your student days and this book offers a chance to see how the landscapes have held up (hint: likely amazingly well).

That is the setup—let us begin the journey.

ORGANIZATION OF THE BOOK

The text begins with an argument for why it is worthwhile to launch an in-depth tour of Walker's built works in the first place. Chapter one also offers up at the outset (no need to bury the lede) what we can recognize as the key, overarching lessons to be drawn from this portfolio. This introduction presents the origins of this study through an anecdote involving a security guard, a molecular collider, and a misunderstanding.

Chapter two gives an overview of Walker's career, illustrating distinct phases while providing an essential timeline for better understanding his professional growth and trajectory. Someone's career is something of a project in itself, and with our detachment and the benefit of hindsight we can recognize some profound lessons within these six decades of design activity. As the rest of the book intentionally does not follow a chronological sequence, chapter two is especially important in defining a temporal structure to which project-related discussions may later be attached.

Chapters three through seven make up the core body of this text, clustered around landscape components of topography, on-structure landscapes, craft, vegetation, and water. Each chapter includes a general discussion of that component's role within landscape architecture, followed by how it has been employed by Walker across a range of built works. Insights gleaned from these projects are clustered as explicit design strategies.

Chapter eight presents a speculative conclusion linking the work of this quintessentially 20th century designer with the challenges and opportunities faced in our 21st century. Highlights from multiple interviews with Peter Walker himself make up the epilogue, reflecting on **landscape as such and legacy of his works**. Finally, the appendix includes additional project details that might be of interest to the reader, in particular site coordinates that may be typed into a map service of choice and used to make a visit to the discussed landscapes—if not in person, then perhaps at least online.

---

1   From a question and answer session at the conclusion of a public lecture delivered at the University of
    Washington in 2003 as part of a retrospective of Haag's legacy. Haag continued, "If you start following this
    line of thought we'd talk ourselves out of a profession," to which the audience responded with laughter.
    I was in attendance and at the time an intern in Haag's office.

2   From Walker's introduction to *Visible, Invisible: Landscape Works of Reed Hilderbrand*
    (New York: Metropolis Books, 2012), 17.

3   For more on Haag's history and philosophy, see Thaïsa Way, *The Landscape Architecture of Richard Haag:
    From Modern Space to Urban Ecological Design* (Seattle: University of Washington Press, 2015).

4   And conversely, casually disregarded by those who have not experienced the challenges of
    professional practice.

5   A topic still in need of more fully being grappled with by the discipline, advocating for built landscapes
    even while acknowledging they often perform less than essential roles.

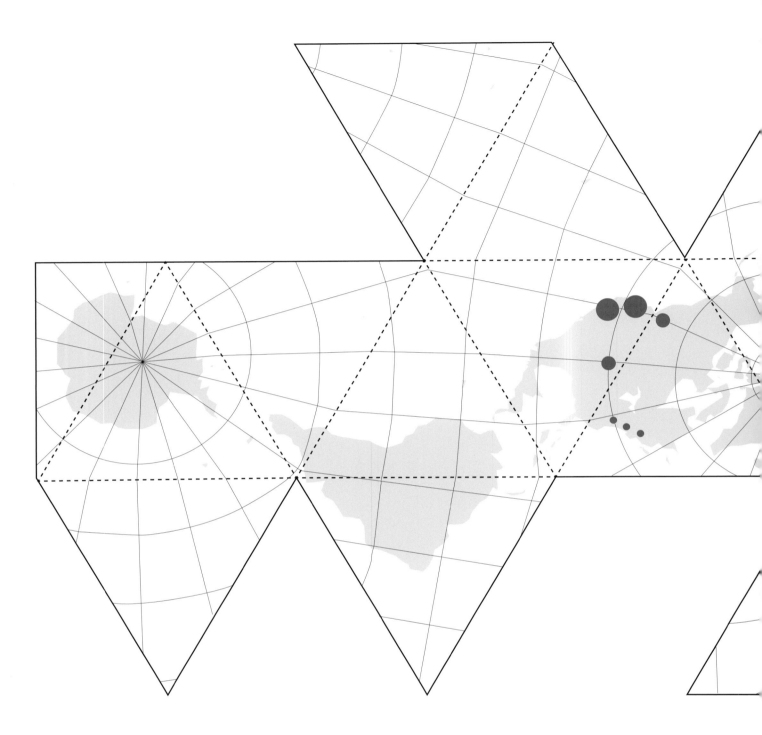

NOTABLE PROJECTS

UNITED STATES

EAST COAST

NATIONAL SEPTEMBER 11 MEMORIAL

TANNER FOUNTAIN

SOUTHWEST

BURNETT PARK

NASHER SCULPTURE CENTER

WEST COAST

FOOTHILL COLLEGE

JAMISON SQUARE

WEYERHAEUSER HEADQUARTERS

The majority of Walker's projects are concentrated on the west coast of the United States, in addition to notable works in other parts of the country.
During the latter half of his career he was also solicited for landscapes in Europe, East Asia, and Australia.

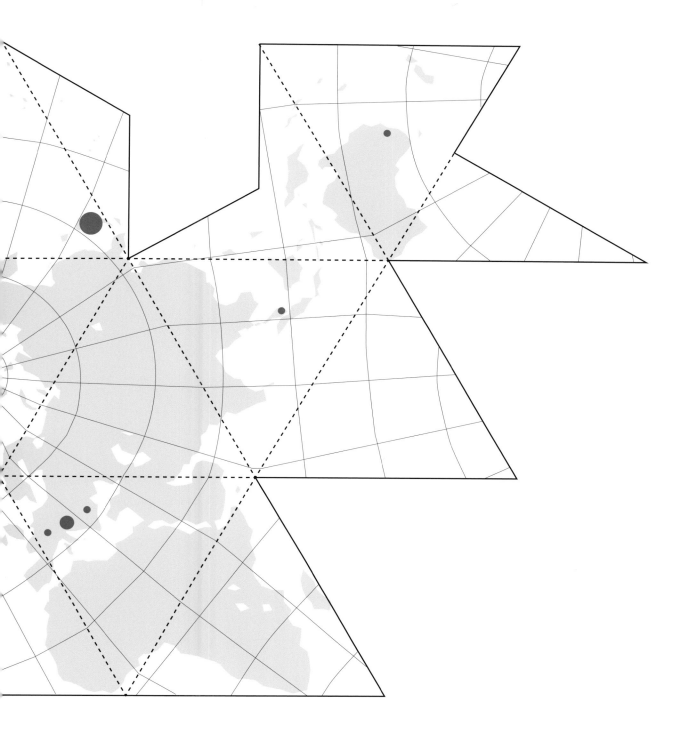

GERMANY

SONY CENTER

SWITZERLAND

NOVARTIS CAMPUS

AUSTRALIA

BARANGAROO

SINGAPORE

MARINA BAY SANDS

JAPAN

CENTER FOR ADVANCED SCIENCE AND TECHNOLOGY

IBM JAPAN MAKUHARI BUILDING

TOYOTA MUNICIPAL MUSEUM OF ART

I

INTRODUCTION

"PETER WALKER!"

The friendly security guard patiently tried to piece together just how our group had arrived unannounced in this otherwise empty entry lounge for the Center for Advanced Science and Technology (CAST), a research facility tucked away in the hills of Hyogo Prefecture, Japan. We would later learn that the guard assumed we were a group of visiting physicists (alas), when in fact this was a dozen landscape architecture students being led by myself and a colleague with an interest in exploring the award-winning landscape of this campus.

Finally, belatedly, we thought to pull up an image of the especially photogenic courtyard on a phone.

"Oh!"

Suddenly there was rapid movement as the guard exited the lobby and gestured to have us follow him into the adjacent building. In front of a large window framing a view of our target, the guard pointed to the courtyard and exclaimed, "Peter Walker!"

Before we all had finished taking our photos he was on the move again, and as we rounded the corner to look through a second window into this rectilinear space, there was the gesture and declaration, "Peter Walker!" Nothing if not thorough, the guard later delivered us to the final viewpoint to this outdoor room and repeated his exclamation, a bright smile of satisfaction on his face.

Most landscape architects have little choice but to come to terms with maintaining a degree of anonymity, as our work is not readily identified and few members of the general public could name a single designer. But on that sunny autumn afternoon within the heart of the sleepy research campus, I was struck by how this facility manager was not only proud of this built landscape but so enthusiastic in declaring the name of its designer—and even more, some two decades after the project's completion. Precious few landscape architects will be so honored.

WHY PETER WALKER, WHY NOW

I am fortunate to work at a university with a well-stocked central library. Roaming the shelves to hunt down sources and gather materials for research work, it is simple enough to experience a mixed sensation of optimism and dread: optimism brought by the great wealth of knowledge available, just waiting to be dusted off, checked out, and enjoyed; dread because of what can be a challenge in identifying where to add one's point of light within this constellation. With so much already written, what more needs to be said?

Well, within the discipline of landscape architecture there is a great deal remaining to be explored, intellectually considered, and effectively conveyed. Ours is a young discipline, rich in history, to be sure, and with an outsized influence on our communities, but young nevertheless. As a modern (lower case "m") profession, we have a modest three or so generations of practitioners from whom to reference, investigate, and learn. Even as landscape architecture is currently experiencing an expansion of visibility and scope through increasingly ambitious public works, we still are challenged to advance beyond anecdotal histories and personalized experience. We are still maturing.[1]

A critical review of a collected body of work offers the opportunity to move beyond "one off" individual observations to instead identify more thoroughly tested strategies. Different forms of groupings bring their own benefits: focusing on a certain geography might reflect contextual cultural conditions; works from a specified window of time can relate to the political economy of that era; grouping by typology (i.e. park, campus, etc.) can illustrate evolving expectations and roles played by those project types. These all are valuable options for structuring design scholarship, but if especially concerned about the role of *designers* and lessons to be learned for integrating the different elements of landscape, the most valuable project grouping will be by individual—we must take on a portfolio.

With the goal of identifying distinct design lessons through a study that is one part synthesis and another distillation, the choice of landscape architect to be considered becomes clear: it must be Peter Walker.

Let us briefly consider Walker's curriculum vitae of accomplishments. Honor Award winner from the American Society of Landscape Architects (ASLA), the highest honor a living landscape architect may be awarded by the organization. Recipient of Harvard's Centennial Medal, the University of Virginia's Thomas Jefferson Medal, and the International Federation of Landscape Architect's Sir Geoffrey Jellicoe Gold Medal. A graduate of the Harvard University Graduate School of Design (GSD), where he returned to serve as professor and chair to the Department of Landscape Architecture in addition to at one time be acting director of the Urban Design Program. He has fulfilled varying roles of teacher, mentor, and collaborator to some of the most prominent practitioners in the field. Co-founder of Sasaki, Walker and Associates, a trailblazing multi-disciplinary firm that continues to maintain international influence in its current form as SWA. An author and publisher, his efforts with Spacemaker Press and magazine *Land Forum* in addition to other publications helped elevate both the visibility and discourse of landscape architecture.

To say that Walker has maintained a conspicuous disciplinary presence, especially within the American context, is an understatement. But even more valuable for our concern, he has remained distinctly prolific throughout this extended career, producing a steady stream of built works not just during one particular phase but in fact spanning a very long time. This then forms the second part of our response as to why Walker—there simply is so much material with which to work.

Plus, we now have the opportunity to view this portfolio *in toto*, and while the vast majority of these projects are (thankfully, remarkably) still in good physical condition, nothing should be taken for granted. Even more significantly, we can recognize a current point of transition for the discipline of landscape architecture that allows us to contextualize these studied works with a bit more distance and, perhaps, even greater appreciation.

ASSESSING A SINGULAR PORTFOLIO

Walker's portfolio offers a singular resource for learning because of three main characteristics: the extensive time span of its development; the sheer volume of projects included; and the exhibited focus maintained throughout these efforts.

Walker entered practice in the 1950s and continued with project work well into the 2010s. While other practitioners have continued their work well past an expected age of retirement—in fact, this would seem to be an especially notable trend within landscape architecture—Walker not only has remained active into his eighties but also began his career in full force while just in his twenties. This extensive timeline means we have projects available for investigation that were produced within a range of contexts, including: post-WWII economic boom of California; emergence of modern environmental movement; contemporary solidification of corporate power; post-modernism and its aftermath; on to expansive globalization fueled by information technologies. Looking to see how much the project work does or does not respond to these contexts proves to be revealing.

The extensive time span is not populated with a sparse scattering of projects, but rather a multitude of landscapes resulting from a consistently productive pace of professional output. And these are not *paper* projects, living only as ideas (although there are an important few of those)—these are built works that survived the development process and are out in the world, ready to be experienced. This level of productivity might on the face of it seem conventional considering that landscape architects are design professionals and by definition need to take on a steady stream of clients and projects. That is to say, these are *practitioners* not artists. Even so, in Walker's portfolio we see a high "hit rate" of projects going through to construction, fulfilled by someone who has embraced an attitude of learning through building.

Across these decades of productive effort, there has been a third defining component—a distinctive degree of focus and iterative innovation woven throughout a range of phases, project types, and locales. Landscape architects, like building architects and other designers, benefit from accumulated knowledge gathered through experience. This is one reason why it can often (but not always) take many years before an individual may reach her or his most productive phase. This learning with time is common enough, but in Walker's work we see a heightened degree of focus and embrace of constraints to provide greater clarity on these insights. For example, with specific landscape components such as topography we can witness certain ideas continuing to be tested and adapted through Walker's first dozen built works. This focus can also be recognized in his use of landscape materials, adapting and reusing specific elements through a sequence of projects. At the same time, Walker's grounding in modernism and later embrace of minimalism only reinforced these methods of focused iteration. With formal clarity and elemental use of materials, the essence and limitations of constructed landscapes are consistently embraced.

In landscape, the transect is used as an analytical device to cut through a terrain and highlight relationships between adjacent elements. In a similar fashion, we have the opportunity to consider Walker's productive half century of practice through a kind of temporal transect revealing the state of landscape architecture as a discipline from the post-WWII years through to the beginnings of the 21st century. Even as we learn from this individual's own journey we have the chance to gain insight on the dynamic contexts within which design is being implemented, offering if nothing else a reminder that present conditions are certain to change.

SITE DESIGN—DOES IT MATTER?

At a time when the dire consequences of climate change linger, increased wealth disparity holds a divisive pull on communities, dramatic urbanization tests the carrying capacity of regional landscapes while creating new challenges of livability within the city core, does something as modest as site design even really matter? The question has to be asked because what follows is an in-depth interpretation for a body of work that, in the end, is all about built sites, ranging from 1 to 1000 acres. The studied material is not a collection of landscape strategies looking to directly take on environmental challenges, or advance ideas of resiliency and adaptation. In fact, most of these projects are at least on the face of it rather static in so much as is possible with any landscape.

With this kind of background, how relevant can the projects possibly be for present and future challenges? What makes them worthy of investigation? Why should anyone care?

Most importantly, sites operate at a scale somewhere between the individual and a community. They are where we get together, and at their best they are civic. Sites are the inverse of abstract—they are perceptible, identifiable, intelligible. Sites can be valuably mundane in the form of a neighborhood park, regional sports complex, or local school grounds. Sites can be aspirational, as the setting for museums or grounds of a university campus. Yes, they can be exclusive as corporate headquarters or private residences—but sites can also be sacred, as places of worship or memorials for gathering in grief and remembrance.

Sites can also be built—they can be *created*. But to do so well, in a way that maintains a high level of physical integrity while also effectively supporting a range of uses over time is highly challenging. It is in this spirit that the investigation is launched, seeing these works not as models to be directly imitated let alone repeated, but rather to be valued and learned from by a next generation of designers finding their own way.

WORD ON CREDIT AND COLLABORATION

Works of landscape architecture demand collaboration, and the examined projects that follow are no exception. So while "Peter Walker" is included in this book's title and all of the following projects are selected from his portfolio, these works were not developed in a vacuum or completed independently by an *auteur*. Each was realized by a team of individuals—a mix of clients, contractors, junior designers, business partners, technical specialists, and others—working together and fulfilling their individual roles.

Some of these landscapes can be recognized as clearly being very personal to Walker himself, while others had significant design efforts led by staff within Walker's studio. The design works included for in-depth study were selected for the clarity and directness with which they might illustrate relevant design lessons.

---

1 For further exploration on this idea of maturing see: R.D. Brown and R.C. Corry (2011), "Evidence-based landscape architecture: The maturing of a profession," *Landscape and Urban Planning*, 100(4), 327–329.

II

THE CAREER
OF
PETER WALKER:
A BRIEF HISTORY

A career can be considered a case study in itself—a kind of *uber*project—offering insights regarding how skills are developed, tested, and advanced over time. Of particular interest in the case of Walker's career is the way in which early successes were built upon in later work, but not in an entirely linear or direct manner. Strategic decisions made along the way and the resulting impacts of those choices are also notable, ready to be interpreted with the benefit of hindsight. There are triumphs, and to be sure some challenges encountered along the way, but more than anything the career described in what follows can be recognized for being especially deliberate with a steadfast pursuit of design and the potential of landscape architecture.

The following brief review of Walker's career also provides a chronological structure of education, firm arrangements, living locales, etc. to which later discussed project works may be attached. So beyond the lessons embedded within this career history itself, establishing an essential timeline will allow us to take a more discrete and elements-focused approach to exploring projects within the core chapters of this study.

KEY DATES

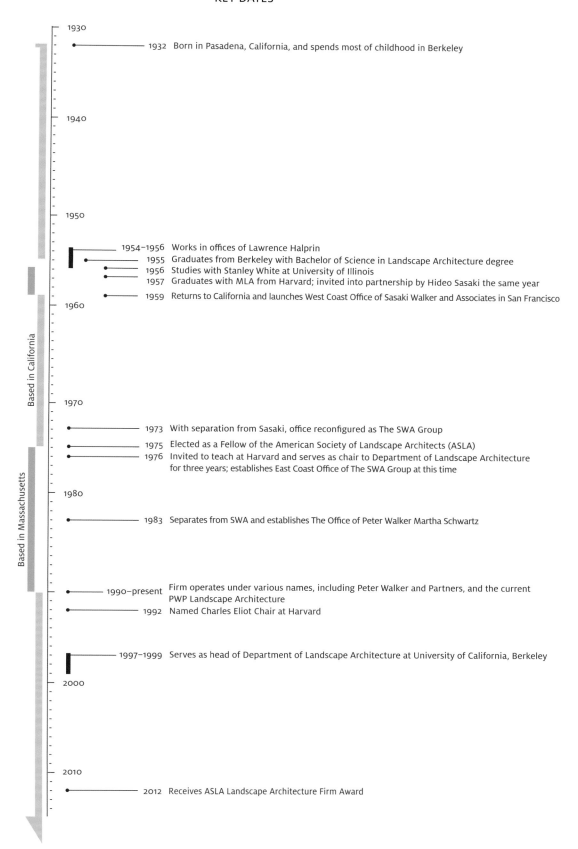

1930

1932 Born in Pasadena, California, and spends most of childhood in Berkeley

1940

1950

1954–1956 Works in offices of Lawrence Halprin
1955 Graduates from Berkeley with Bachelor of Science in Landscape Architecture degree
1956 Studies with Stanley White at University of Illinois
1957 Graduates with MLA from Harvard; invited into partnership by Hideo Sasaki the same year
1959 Returns to California and launches West Coast Office of Sasaki Walker and Associates in San Francisco

1960

Based in California

1970

1973 With separation from Sasaki, office reconfigured as The SWA Group

1975 Elected as a Fellow of the American Society of Landscape Architects (ASLA)

1976 Invited to teach at Harvard and serves as chair to Department of Landscape Architecture
for three years; establishes East Coast Office of The SWA Group at this time

1980

1983 Separates from SWA and establishes The Office of Peter Walker Martha Schwartz

Based in Massachusetts

1990–present Firm operates under various names, including Peter Walker and Partners, and the current
PWP Landscape Architecture
1992 Named Charles Eliot Chair at Harvard

1997–1999 Serves as head of Department of Landscape Architecture at University of California, Berkeley

2000

2010

2012 Receives ASLA Landscape Architecture Firm Award

Peter Walker is a Californian.

Born in Pasadena (a city within Los Angeles County) in 1932, Walker's early childhood coincided with when the United States was still in the grips of its Great Depression. After the death of his father, a young Peter Walker moved with his mother and stepfather to Berkeley, a small city east of San Francisco. More than any other locale, California's Bay Area generally and Berkeley in particular would come to be the home base for the majority of Walker's life and career.

After graduating from Berkeley High School, it was at nearby University of California at Berkeley that Walker completed his undergraduate studies in landscape architecture. During this time Walker also gained some of his first professional experience, working from 1954 to 1956 in the office of renowned San Francisco-based landscape architect Lawrence (Larry) Halprin.

The young Walker traversed[1] between Northern and Southern California during his childhood years. In this era, before the onset of commercial aviation, hours on rail and road offered a chance to gain expansive knowledge about the state's diverse geography. This understanding and general rootedness would prove to serve him well, for even while ultimately holding international influence, the majority of his projects for the following years would be located in California.

After graduating with his bachelor in landscape architecture degree in 1955, Walker moved to the University of Illinois to begin his graduate studies. At Illinois, Walker studied with and worked for Stanley (Stan) White, a key encounter with someone who held notable influence on other young designers of this era.[2] At the suggestion of Stan White, Walker continued his studies at Harvard University Graduate School of Design where he graduated with his master in landscape architecture degree in 1957 and met Hideo Sasaki.

MENTORSHIP TURNED PARTNERSHIP WITH SASAKI

There would be no "Peter Walker" as we have come to know him without Hideo Sasaki.

Hideo Sasaki completed his MLA studies at Harvard Graduate School of Design in 1948,[3] when Walter Gropius was still in command of the school's architecture department and modernist ideals were in fresh circulation. During his decade-long tenure as chair of the school's landscape architecture department, Sasaki facilitated an open exchange of knowledge between his office and the school, augmenting increased teaching needs (especially in regard to design) by supplying his own staff while at the same time taking advantage of the opportunity to hire bright graduates. It was in this setting that Walker joined Sasaki's practice based in Watertown (a short distance from the Harvard campus) and was made partner in 1957; Walker was still only in his mid-twenties.

In hindsight, we can recognize a mix of commonalities and differences that helped make Sasaki and Walker such an effective pair. Both men were born and raised in California,[4] and both studied under Stan White at the University of Illinois. Sasaki's legacy is in valuing and effectively managing cross-disciplinary collaboration through project works that often mixed planning and design to great success, while with Walker we can recognize a more distinct design voice present in the built works. This compatibility on process and a complementing set of skills was further emphasized by their age difference (Sasaki was born in 1919), meaning that one could make the most of this young talent while the other could benefit from his partner's experience and earned prestige.

Two years after establishing their partnership, Walker returned to California to establish the West Coast Office of Sasaki, Walker and Associates. It was ongoing work for Foothill College and the requirement of a local field office that initially triggered this move, but with the post-World War II building boom fully underway the firm was entering into what Walker has described as a "supercharged development environment."[5] The firm would go on to execute a range of works, including: campus planning and design, mixed-use developments, and what at the time was described as urban renewal projects. It was a period of growth for not only the region but also the firm itself. When Walker joined Sasaki in Watertown the office had a modest staff of six or so, whereas a decade later they had over two hundred professionals within the firm. Beginning with the core group of landscape architects, the office became increasingly multi-disciplinary over time with the inclusion of ecologists, civil engineers, and other specialists.

The conscious embrace of modernism, the testing of its potential within the medium of landscape, and tensions regarding the legibility of this design approach for works realized through the atelier versus more corporate firm structures, all were explored during this critical phase of Walker's career. When Walker returned to California after his studies at Harvard and partnership with Sasaki, he was rejoining a community of landscape architects developing projects ambitious for their experimentation and testing of new approaches to landscape design, even if the scale of these projects in the late 1950s and early 1960s was quite modest relative to what would later be built. Active San Francisco-based designers of the time included Garrett Eckbo, Thomas Church, and Lawrence Halprin, along with Richard Haag before his departure for Seattle.[6]

All of these designers would in time come to be recognized as leaders in advancing modernism in landscape architecture. External conditions were amenable to these projects, as postwar development was gaining momentum (but, importantly, was not yet at its zenith) and creating significant demand for design talent, while the shifting expectations of a growing middle class heightened focus on outdoor spaces that supported leisure and the nuclear family. At the same time, California's climate supported year-round outdoor activity while the less conspicuous presence of cultural institutions translated into fewer preconceived standards.[7] These factors together made the west coast especially inviting.

Within this context, Walker was navigating shifting interests in design culture generally, and landscape architecture in particular. In his essay "The Practice of Landscape Architecture in the Postwar United States," he begins by describing the prevailing sentiment during his early studies. "In 1955, when I graduated from Berkeley, we were told, primarily by our architectural professors, that the past was not only dead but useless in our quest for modernist beauty."[8] A generally unapologetic embrace of modernism, its aesthetic language and goals of universality at the time, is described with regards to the Deere Headquarters in *Invisible Gardens* (written by Walker and Melanie Simo), where they write of how "[a] postmodern preoccupation with context, metaphor, and iconography may discourage them [critics of today] from trying to see the Deere Headquarters in its own time—a time when modernism encouraged dreams of pure form, form with a life of its own, and universal space."[9] The potential of modernist approaches and the related (but distinct) clarity of minimalist works would continue to play a central role for the rest of Walker's career.

These early years with Sasaki also brought a range of influential attitudes: confidence in pursuing projects of large scope and scale; a general *modesty* for how landscape often must relate to architecture; an emphasis on asking good questions of stakeholders and identifying key issues; the value of collaboration and working as a team. The lasting impact of these attitudes would continue to be apparent through Walker's own work across the decades, even as he clearly found his own voice and developed his own distinct set of priorities.

CORPORATE PHASE WITH SWA

The incessant stream of projects completed during the 1960s and on into the 1970s provided Walker with a potent experience base to build upon, but establishing this foundation was highly demanding. It was created via a way of working from which Walker eventually stepped away.

Projects from this era include urban renewal efforts like Sydney Walton Square and the adjacent Alcoa Plaza (completed between 1960–1968) in San Francisco, the latter of which was especially notable as the firm's first fully on-structure landscape. There also were corporate facilities, such as Security Pacific National Bank (1974) and the beguiling Weyerhaeuser Headquarters (1972). New town developments built to accommodate California's growing population were also prevalent, most markedly a whole collection of neighborhood, civic, and commercial works for the Irvine Company in the southern half of the state. Although Walker would later make a very conscious choice in no longer pursuing such development-driven projects, it is striking to recognize just how many of these landscapes are still in place and exhibiting a robust physical condition. An inspired form of pragmatism that worked with rather than resisted the limits of landscape would seem to be at the heart of this success—an approach further explored in the chapters that follow.

The firm's structuring proved effective at absorbing continued growth and increased disciplinary diversity, although this lessening of focus also at times caused internal conflict. The formal separation between Sasaki Walker West and Sasaki Walker East occurred once the east coast branch began offering in-house architectural services, drawing concern this might limit opportunities for the west coast office and their collaboration with a wider range of architects. With the separation between east and west offices complete, the California-based firm was officially renamed The SWA Group, with Walker serving as chairman.

Key characteristics of The SWA Group put in place at this time still define the firm (currently known simply as SWA) to the present day. This includes its team-driven structuring, with a lead design principal paired up with an associate in charge of project management for each project. The general clarity of responsibilities can be effective in external communications with clients while accommodating other changes to the design team as deemed necessary during the life of the project. The employee stock ownership plan (ESOP) put in place during these still early days of The SWA Group has had an even more lasting impact. One of the first companies in the country to implement such a system, the ESOP provides a means for employees to become increasingly vested in the company. The net result is a hierarchy flatter than that typical of other private firms, with highly skilled and relatively long-term associates playing a central role.

The first major shock to The SWA Group's growth arrived in 1973, with that year's oil embargo and consequential recession. Land speculation and related development-driven work came to a standstill, and the firm laid off half of its staff. It was a challenging time. "After The SWA Group recovered from the recession I was personally exhausted,"[10] Walker reflects. By 1976, he was ready for a change and even considered dropping landscape architecture altogether to become an artist.[11] It was at this pivotal moment that he was contacted by Charles Harris, chair of the GSD's Department of Landscape Architecture from 1968 to 1978—asking if Walker would return to Harvard.

HARVARD, ART, AND MARTHA SCHWARTZ

When Walker made a return to Cambridge, Massachusetts, in 1976, it was as a teacher rather than student, launching a phase of introspection and exploration that would define his further career.

Initially serving as acting director for the Urban Design Program, he went on to be chair of the Harvard Graduate School of Design (GSD) Landscape Architecture Department from 1978 to 1981 before eventually being named Charles Eliot Chair in 1992. Walker had been called upon to help implement a more design-focused curriculum, which he carried out by drawing from his exhaustive professional experiences gathered during the previous two decades while also relating interests in contemporary art and its potential relationship to landscape design. A specific and especially lasting pedagogical legacy from this time was the innovative use of model making, especially during early studio works. Often making use of unsophisticated everyday materials such as cardboard and foam, the process of making these models forced a distillation of design ideas while also producing an especially engaging form of representation. Multiple generations of GSD students have by now been introduced to this method, some making great use of it in their own professional work upon graduation while others have disseminated the practice through professorship roles across the globe.

Art, and minimalist art in particular, gave inspiration and provided its own kind of formal guidance to Walker during this period. Interest in this area had started earlier while accumulating a growing personal collection of significant contemporary artworks, initially with no direct relation to his professional work. Right as the rest of design culture began to shift toward a fascination with post-modernism, Walker together with gallerist Harold Reed assembled an exhibition titled "Art Into Landscape" that was hosted in Reed's New York gallery in 1980. A review of the event's brochure makes clear the ambition— the *sentiment*—behind the exhibition, with the document's succinct introduction offering that "[t]he current estrangement of the art world and the world of design is peculiar given the pronouncements of the modernists for social, collaborative and artistic opportunities in industrial and post-industrial societies. In the exhibit we hope to briefly pose a hypothesis of explanation and suggest a cause for hope."[12]

Across five spreads of the brochure, a clustering of images depicting sculptures or landscapes is paired with text describing ways in which such pieces have expanded expectations of garden spaces. While early pages include works by artists such as Henry Moore, Joan Miró, and Carl Andre, the middle spread transitions to works by Nancy Holt, Christo and Jeanne-Claude, Robert Smithson, and others depicting engagement with broader landscapes through discrete installations. The booklet concludes with some of Walker's more speculative work at the time, such as a non-vegetated roof garden installation from 1979, marking a deceptively important moment: this was to be one of his last works under the auspices of The SWA Group and the first developed in collaboration with Martha Schwartz.

Not long before this exhibit, Walker had given serious thought to departing from the discipline of landscape architecture to more fully pursue art itself. Even the possibility of such a change can be something of a shock in retrospect, but at the same time offers telling insight into his own ambition. Evidence of lingering interest in this repositioning can be seen in brochure captions accompanying the sample works that include dimensions (i.e. 20 × 60 feet or approximately 6.1 × 18.3 meters), much as would be done with other artworks. Still in his forties, at this moment Walker was holding a leadership position for one of landscape architecture's most successful firms yet organizing intimate art-focused exhibits; he was department chair at Harvard, and yet considering a dramatic career change.

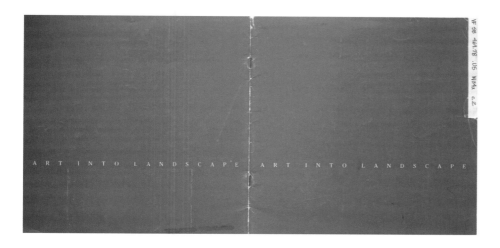

ART INTO LANDSCAPE—At present there are signs in architecture and landscape architecture that the functionalist imperative of the Modern Movement is now being balanced with visual and humanistic values. The "post modernists" led by architects Philip Johnson, Charles Moore, Michael Graves, Frank Gehry, Aldo Rossi and others are rekindling the designer's interest in artistic issues. Likewise, artists concerned with environment-building are pushing toward large scale works which allow or imply physical entry.

Perhaps the long-standing need for more intimate collaboration and awareness between the fine arts and design in the landscape may be at hand. We hope that the course of these so long separate activities will continue to extend their current beneficial overlap. No matter how these ideas are synthesized there will continue to be a need for gardens and public spaces that relax the body, inform the mind and refresh the soul.

Peter Walker

Chairman of the Department of Landscape Architecture
Harvard Graduate School of Design

SWA GROUP LANDSCAPE ARCHITECTS
PETER WALKER, MARTHA SCHWARTZ, JOHN WONG
DESIGNERS
"A Roof Garden for Art Collectors
in Back Bay, Boston"
Gravel, Astro-turf, Aluminum tread-plate,
Flower-pots, Wood
22' x 60'
1979

Brochure for "Art Into Landscape", an exhibition organized by Peter Walker and New York gallerist Harold Reed in 1980.

To his ultimate success and the discipline's benefit, instead of departing from landscape Walker instead worked toward synthesizing his interests in art and minimalism with "the craft I had learned over the last 20 years."[13] He departed from The SWA Group and formed a personal and professional partnership with Martha Schwartz. As experimentation such as the roof gardens and then the Necco Gardens (1980)—a temporary installation overlaying two shifted grids to define space and test expectations of formal gardens on the MIT Campus—continued to grow in ambition, the decision was made to establish a new practice.

Working together as The Office of Peter Walker Martha Schwartz, the two had the everyday challenges of establishing and managing a staffed office in addition to successfully relating their own budding design interests with the expectations of their clients. In Walker's telling it was simple enough to make use of his professional network to find initial work, but then the clients were sometimes unpleasantly surprised by what was being offered. "And I'd never been fired before and when they'd find out what we were trying to do, they'd fire us. So that was different, even though we were small, five, six, seven people."[14] The decision was made to seek out the kinds of clients who might be supportive of their work in particular, rather than clients looking to be supplied with more general professional services. "I wanted institutions because I wanted to build well and I wanted someone who would take care of it."[15] Corporations, universities, museums, and municipalities were to be the primary focus for the next three decades.

FOCUSED PARTNERSHIPS

Walker and Schwartz were determined to return to California, setting up shop in San Francisco as a kind of homecoming for Walker. After the two separated first professionally and then personally (Schwartz went on to run a successful practice based in London and New York), Walker's firm went through a number of permutations: Peter Walker William Johnson and Partners, Peter Walker and Partners Landscape Architecture, and on to the current PWP Landscape Architecture. This latter phase of Walker's career proved to be highly fruitful and resulted in his most accomplished works.

The plan to pursue institutional clients was, in a word, successful. An especially important client was IBM. Planning for the firm's Solana Campus in Texas (1989), a 900-acre (364-hectare) development for some 20,000 employees, began in the mid-1980s and triggered a growth phase for the firm. IBM brought related projects in California, and later with their IBM Japan Makuhari Building (1991) facility outside of Tokyo created a point of entry for the firm's work in Japan. Universities have been reliable partners as well, with Stanford University alone providing half a dozen projects. There have been a variety of public parks and plazas spread out across the west coast. Museum landscapes as exemplified at the Toyota Municipal Museum of Art (1995) and Nasher Sculpture Center (2003), present a high watermark for build quality that surpasses corporate projects while still being accessible to the public. And with the National September 11 Memorial (2011), a lifetime of knowledge and expertise was employed to negotiate tremendous complexity in creating a landscape of remembrance.

Institutions as clients have been one important component, but successful collaboration with (often high-profile) architects has been another. This has included the likes of Yoshio Taniguchi, Renzo Piano, Foster and Partners, and SOM. Meanwhile, with Walker's past experience at SWA and experience-gained knowledge regarding the pros and cons of different studio structures, a conscious decision was made to not have branch offices while also limiting the firm to thirty or so professionals. These self-imposed limitations have allowed for maintaining focus and oversight across the years.

There is much more that can be said about this history—indeed, the legacy of junior staff members who have gone on to lead their own successful firms could itself be a rich topic of discussion. But for our purposes it is the legacy of the projects themselves and the lessons to be drawn from their physical condition that is of greatest interest.

As for PWP today, the firm is entering into a new phase as it builds upon its distinct foundation. Design firms with a singular leader often disappear once that individual withdraws from day-to-day involvement, but that clearly is not the case with PWP. The challenges for today's PWP are shared with other contemporary practices: how to be responsive to an increasingly dynamic environment; implement ideas of craft at increased scale and speed; remain rooted in place even while operating at a global reach; and ultimately, make use of the elements of landscape to shape spaces that have value, have meaning.

An in-depth exploration of selected landscapes can, it is hoped, help inform responses to these critical questions.

[1]   This and other valuable biographical details are described in The Cultural Landscape Foundation's "Pioneers of American Landscape Design" series: https://tclf.org/pioneer/peter-walker

[2]   White's roster of influenced and influential design students included Richard Haag, Charles Harris, and Stuart Dawson, in addition to Hideo Sasaki and Peter Walker. Stanley White was brother of noted author E. B. White and earlier in his career worked in the Olmsted Brothers office.

[3]   For an in-depth study of Sasaki's life and legacy see Melanie Simo, *The Offices of Hideo Sasaki : A Corporate History* (Berkeley, CA: Spacemaker Press, 2001).

[4]   At the onset of World War II, Sasaki spent a brief time at an internment camp for Japanese Americans before leaving with his family for Denver, where the governor of Colorado welcomed those who had been evicted.

[5]   See Peter Walker, *PWP Landscape Architecture: Building Ideas* (San Rafael, CA: Oro Editions, 2016), 16.

[6]   Haag was a close friend and former student of Sasaki's, and it was from a rented desk within Haag's office that Walker initially launched the west coast branch of Sasaki, Walker and Associates.

[7]   For more on this period see Mark Treib, ed., *Modern Landscape Architecture: A Critical Review* (Cambridge, MA: MIT Press, 1993).

[8]   Peter Walker, "The Practice of Landscape Architecture in the Postwar United States," in Mark Treib, ed., *Modern Landscape Architecture: A Critical Review* (Cambridge, MA: MIT Press, 1993), 250.

[9]   Peter Walker and Melanie Simo, *Invisible Gardens: The Search for Modernism in the American Landscape* (Cambridge, MA: MIT Press, 1996), 238.

[10] Peter Walker, *PWP Landscape Architecture: Building Ideas* (San Rafael, CA: Oro Editions, 2016), 21.

[11] To this end, Walker even obtained a studio in New York's SoHo district.

[12] Peter Walker and Harold Reed (1980), *Art Into Landscape: An exhibit organized by Peter Walker in conjunction with Harold Reed*; April 29–May 24, 1980. New York: Harold Reed Gallery, 1.

[13] Peter Walker, *PWP Landscape Architecture: Building Ideas* (San Rafael, CA: Oro Editions, 2016), 21.

[14] From interview with author, Berkeley, 29 May 2014.

[15] From interview with author, Berkeley, 29 May 2014.

LANDSCAPE COMPONENTS

Every landscape, be it built or naturally occurring, is a composite of inert and living materials that interact with the surrounding environment. There is the materiality of the geology itself providing structure, a skin-like layer of living soils on the surface that nurtures vegetation, and a constant cycle of water falling as precipitation, partially absorbed by the sponge of landscape and partially collecting as surface runoff, the infiltrated portion of which can eventually be returned to the atmosphere via evapotranspiration.

In creating *built* landscapes, these natural elements and processes are reconfigured in artificial arrangements to achieve desired outcomes. With finished works often (but not always) obscuring the complexity of these systems it can take a keen eye to unpack the distinct landscape layers. This effort of deconstructing built landscapes can be challenging, but also rewarding in providing new insights.

For Walker's portfolio in particular, what lenses of seeing ought to be used for best identifying specific design strategies embedded within these works?

We begin with **topography**, the most essential landscape component for shaping the land. In Walker's work there is a general divide between his earlier and later projects; initially he makes use of fine-scaled surface modulations that are highly inviting but that he eventually considers at risk of being too identifiable and pastoral. In later works there is a greater fascination with flatness and the technical challenges in achieving this effect with living materials. All the while, existing topography is interpreted and interacted with in particular on projects that include sited buildings.

On-structure landscapes—that is, landscapes built upon rooftops or some other form of building structure—are both demanding and vulnerable, typically requiring significant commitment of engineering and other technical needs while the perched vegetation must grow in constrained conditions that are also more highly exposed than what would be experienced on the ground. In Walker's portfolio we have the great advantage of a wide range of on-structure landscapes to learn from, from early works that are technically relatively simple but also highly robust even after fifty years since their creation, all the way up to more recent projects where on-structure knowledge has allowed for increasingly extensive and sophisticated environments.

Landscape presents distinct challenges for fulfilling ideals of **craft**, but this ongoing concern for Walker has resulted in iterative developments for working within material and performance limits. Some of these projects, especially from a later phase of his career, do in fact represent especially high-end works within the discipline where construction budgets were far above the norm. But for still even more projects, craft is represented more in the thinking and careful consideration for how elements are fitted together, irrespective of cost. These examples of modest materials that have been thoughtfully employed provide especially rich insights.

For many, **vegetation** is nearly synonymous with landscape, and although Walker has experimented with the potential of vegetation-free landscapes at pivotal moments in his career, plantings overall play tremendously important roles in his projects. And that is roles, plural—when experiencing and then critically interpreting these projects, the specific goals and impacts of these planting choices are revealed. The net result is a kind of toolkit for relating vegetation with site conditions and intended experiences.

Our systematic investigation of Walker's portfolio concludes with **water**. This inherently dynamic material accentuates "moments" in a built landscape through its animated qualities, and yet the technical and consequential financial requirements for constructing robust water features mean they get built far less often than originally intended. Even so, across Walker's extensive collection of projects there is a notable consistency in water features he managed to include in the finished work. How was this accomplished, and to what end? These two questions are to be held in mind across our exploration.

Nishi Harima Science Garden City, 1993. [Photographed 18 years after construction]

III

TOPOGRAPHY: SHAPING THE LAND

It begins with the land, the earth itself, *terra firma*. When it comes to reshaping the landscape to what is desired, the manipulation of the ground plane marks the essential—and often, most effective—place to begin. Translating design aspirations to built form demands two related topographic actions: reading and manipulation.

The **reading** is most relevant during the stages of site planning, when decisions regarding the positioning of structures and arrangement of circulation have a lasting impact on any project. When working with multi-disciplinary teams, decisions made in these early stages give landscape architects the opportunity to offer real value to a project while gaining trust from allied colleagues. The act of **manipulating** the ground plane is more projective than responsive, with the designer setting about to sculpt the land in the service of programmatic and experiential goals. In this way the landscape architect is operating as a shaper of space.

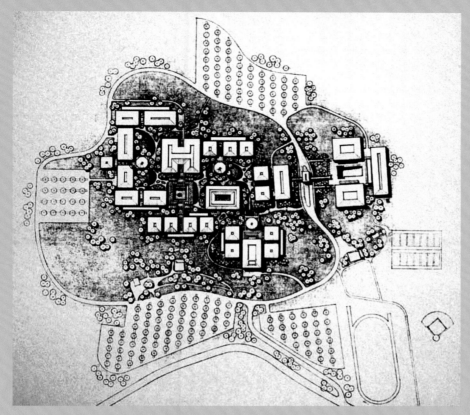

Foothill College. Original campus site plan, with elevated courtyard at center and parking relegated to downslope former orchards.

GUIDING CIRCULATION, MAXIMIZING SPACE

Some of the most immediate impacts of ground plane manipulation include the guidance of pedestrian circulation and subdivision of outdoor spaces. Four of Walker's projects are especially effective in illustrating how different forms and degrees of manipulation bring a range of results.

A sequence of landforms guides pedestrian circulation and defines a sequence of outdoor rooms within **Foothill College**, Los Altos Hills, California, 1957–1960. [Photographed 54 years after construction]

At the heart of **Foothill College** in Los Altos Hills in Northern California (1957–1960) is an expansive sequence of open spaces defined by a collection of low-rise buildings that house classrooms, a library, and other facilities. Pathways linking these structures are serpentine in form, with the asphalt paving spilling out to fill low ground between a number of turf-covered mounds. Moving through the space feels leisurely, with the scale of the landforms just large enough to generally dissuade shortcuts being taken as desire paths across the planted areas, while still gentle enough to be inviting.

The sequence of landforms also effectively subdivides and layers views out across the courtyard spaces. Something like the way a room in a building can actually feel larger with the addition of furniture, the landforms help make these outdoor spaces feel more intimate and less confined. Having the landforms pressed up to the edges of the structures accentuates this effect, as the edges of the space become obfuscated.

Something not readily apparent in the plan drawings or photographs is the way in which this site is perched atop a hill, with users arriving by a steady climb in some form of vehicle and then alighting at the clustered parking and transit lots. From these perimeter areas the heart of the campus is arrived at on foot, requiring a final climb to reach the central open spaces. This sequence of climbing and landform heightens the sense of arrival while giving the space an otherworldly feeling as a room in the

Foothill College. Entry sequence, climbing from lower parking areas to pedestrian bridge.

clouds. "In my first public project as a young landscape architect in 1960 the grading both set the site plan and produced the human scale of the pedestrian campus."[1] Walker's reading of the site's topographic context and referencing this in the more finely scaled landscape design resulted in a multi-scalar experience of landform.

The undulating landforms within **Sydney Walton Square** (1968) are in a similar language to those at Foothill College, but with enough differences to be instructive. Both of these sites fulfill the role as landscape rooms that may be easily traversed or function as destinations in themselves. While both are located in Northern California, Foothill College is perched within suburban Los Altos Hills[2] while Sydney Walton Square is located near the waterfront (and sea level) of downtown San Francisco. Moving through Sydney Walton Square today, it is clear to see how the shifting topography helps provide visual interest while accommodating different modes of use: office workers taking lunch in the outdoors; individuals finding a place of repose within the busy city; or indeed, more actively programmed special events that spill out into the flatter turf areas and host larger groups. The park still successfully supports all of these activities, in addition to providing a platform for multiple works of sculpture, and yet it lacks the timeless quality experienced at Foothill College. Why might this be? Context plays a significant role—detached and in the hills, one can imagine Foothill College's one hundred year anniversary with the bones of the site largely unchanged. Meanwhile, Sydney Walton Square today feels out of place—more than that, out of *sync*—with its context and the many challenges of this urban environment.

Gently undulating topography guiding pedestrian movement at **Sydney Walton Square**, 1968, located in San Francisco's Financial District.
[Photographed 50 years after construction]

Another distinction between the two previous projects is the typical height and general profile of the mounds within the landscape, with the former being a bit more aggressive in steepness and often meeting the paving edge as a convex rather than concave curve. These variables of slope profile were tested through additional configurations in later projects, including the **South Coast Plaza Town Center** in Costa Mesa, California (1991). This site is really an amalgamation of multiple projects fitted within one masterplan executed over many years, mixing commerce, culture, and corporate presence in a distinctly Orange County, California, fashion. A central open space described as the Town Center Park provides a green pedestrian space in this otherwise automobile-dominant area, linking immediately adjacent buildings in addition to accessing a nearby shopping mall reached via a nearby pedestrian bridge. Notably, Isamu Noguchi's majestic sculpture garden California Scenario is located just across the street.

Viewed in plan, the Town Center Park has clear similarities with Foothill College, but when experienced on the ground the impact of seemingly minor differences becomes apparent. For one, the width of the pathways is reduced in the latter project, allowing the planted areas to be dominant and the asphalt paving to play a secondary role, slicing through the space. The landforms are also more pronounced and sculptural in effect, yet less objectified due to topographic change present throughout the ground plane and undulating walkways. The cohesiveness of the space is further strengthened by a much more intensive tree planting scheme that provides greater shade but also a sequence of limbed-up trunks that emphasize the sense of movement when walking through the space. The net result of these design choices is that this modestly scaled and ostensibly rather *simple* landscape is one that is identifiable, a place that is memorable.

The interplay of topography, pedestrian paving, and tree plantings form an inviting and memorable landscape within **South Coast Plaza Town Center**, Costa Mesa, 1991.

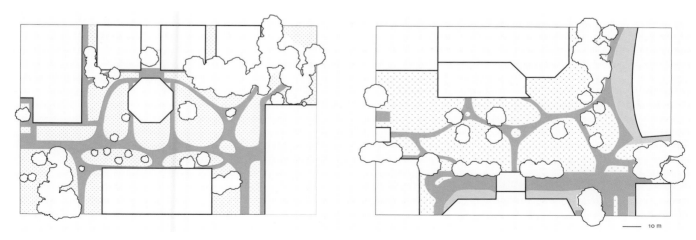

Figure-ground comparison of paving vis-à-vis planted areas at **Foothill College**, 1957–1960 (left)
and **South Coast Plaza Town Center**, 1991 (right), the latter (and later) of which strikes a finer balance.

41

If the topographic formations within South Coast Plaza Town Center represent a mature balancing of elements, the landforms originally part of the **Children's Pond and Park** (1997) offer a more cautionary tale. This park is located near a convention center in downtown San Diego, adjacent to a waterfront thoroughfare also designed by Walker's office and containing a large water feature further discussed in this book's water chapter. The area of the park was originally dominated by an array of mounds uniform in structure and nearly eye-level in height, interspersed with decomposed granite and densely planted pine trees emerging out of this soft paving. The scheme can be read as an extension of ideas tried out at the Town Center Park and earlier works, but more sculptural and, yes, more minimalist.

In the end, however, also considered more dangerous. Early photos show how this configuration of elements had been alluring, with the landforms built at an inviting scale where visitors could sit or even lie down on the grassy slope. At the same time, the formality of these perfectly circular mounds allowed for greater *informality* with the paving, and that in turn supported especially sculptural planting conditions for the trees as they emerged directly out of the decomposed granite. In short, it was a masterful composition and yet ultimately unable to survive the combined enveloping urban challenges of drug use, homelessness, and safety. Eventually the landforms were removed by the local government and now the rings of paving defining their base are all that remain. Walking through the site today, this is a sad place that feels utterly isolated despite its city center location. Beyond the specifics of its design elements, this project offers a reminder for how vulnerable any park can be that does not have some form of ownership by a nearby community.

Turf mounds within the **Children's Pond and Park**, San Diego, viewed soon after construction in 1997 and also more recently, after the removal of these landforms.

Ultimately, Walker stepped away from the kind of sweeping topographic manipulation evident in his early projects, considering them at risk of being overly pastoral and also too easily identifiable. The objectified landforms would continue to be used, but there also was another topographic operation being tested in parallel: flatness.

Zoom out far enough and any landscape begins to appear flat. It is something that junior designers are reminded of when first preparing long sectional drawings where significant landscape features get reduced down to imperceptible bumps along the cut line. And yet in the experience of a place at the human scale, the physicality of landscape makes true flatness nearly impossible to achieve. Soils heave and contract while vegetation changes over time. Even for hardscape areas there are challenges, especially in light of managing stormwater runoff and accommodating base conditions prone to compaction and lateral movement. These combined challenges result in landscape designers often circumventing the use of distinctly flat areas within their designs, but for Walker it has been a subject of focus and ongoing experimentation.

Burnett Park (1983, with revisions in 2010) is scaled and positioned as a quintessential urban green space within the mid-sized city of Fort Worth, Texas. Two acres (approximately 8000 square meters) in size, the rectangular site is bordered by city streets on three sides and an office tower on the fourth. The landscape itself is a composite of distinct layers: mature trees and shrubs, most preexisting from an early iteration of the park and largely positioned at the periphery; flat platforms of turf; and granite pathways elevated above this lawn layer, stitching across the site. Originally, there was a rectangular band of pools that provided yet another layer, but these were removed as part of renovations completed in 2010.

The flatness of the site helps it function as a connecting space, something especially visible as office workers make their way across the site and to the adjacent tower's main entrance that is oriented toward the park. Flatness here also supports functionality and diverse usage of the space, accommodating everything from public festivals to wedding ceremonies. But what elevates the project to something remarkable is the way this flatness is made legible through careful detailing, most notably the edges of the granite paths that provide some space between the stone and turf and with that cast a shadow line.

The project's listing in Walker's first monograph describes how "the care and intricacies of Burnett Park's design create a new awarenesses of space and function."[3] Visiting the park thirty-five years after its creation, these qualities are still on vivid, inspiring display. Burnett Park offers the lesson in how flatness can increase the utility of an urban green space, while the fine detailing used to achieve this flatness increases the landscape's value even as an empty vessel.

Layering of hard and soft elements, casting shadows and expressing flatness at **Burnett Park**, Fort Worth, 1983.
[Photographed 35 years after construction]

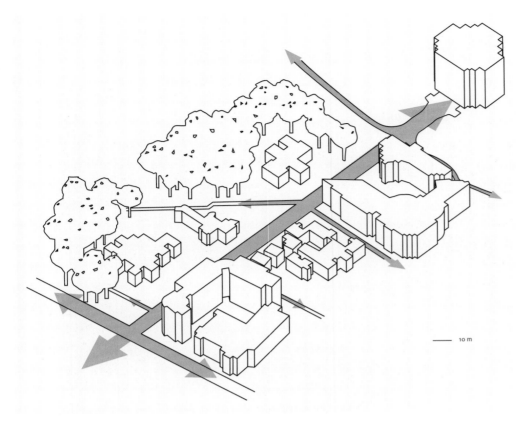

The flatness of the pedestrian axis helps to link disparate pathways while creating a public space in its own right.

Flatness can also be a statement, an intervention made within a larger landscape that functions as a unifying element. At the University of California at **Library Walk**, University of California at San Diego (1995), this intervention takes the form of a level axis that begins with a main access point for the university and terminates at the entrance to an expanded library. Across this span the walk links a variety of campus facilities, creating a pedestrian spine that functions as a conduit, wayfinding element, and destination in itself. In making the axis flat it becomes primary—the datum—amongst surrounding circulation paths.

Essentially a flat, straight sidewalk—if it seems simple, that is because it is, conceptually. But in experience this simple gesture is dramatically effective, especially within a university context. As universities grow over time their campuses can begin to offer uneven experiences, an amalgamation of many individual projects where areas that had previously been considered "back of house" later become more important frontage. The student of design is reminded of this when navigating a mix of access roads and small plazas on the UCSD campus before encountering the Library Walk and its clarity of spatial structure. This form of space-making in San Diego foreshadows a later project at the University of Texas at Dallas, only there water plays a significant role—further explored in the water chapter.

Linear sequence linking campus facilities at University of California at **Library Walk**, University of California at San Diego, 1995. [Photographed 22 years after construction]

At rear of image, use of berm to define space at **Toyota Municipal Museum of Art**, Toyota City, 1995.
[Photographed 17 years after construction]

BERMING AS DEFINITION

Berms provide an efficient and lasting means of spatial definition, distinct from tree massing or site walls. Spread across Walker's portfolio we find a range of examples for where this topographic device has been deployed, often in a strategic manner to mitigate certain constraints of the site.

Most of the landscape at **Toyota Municipal Museum of Art** in Toyota City, Japan (1995) is contained within two terraces, creating platforms for the display of art while building linkages between the museum and nearby historic structures. One conspicuous deviation from this flatness appears at the upper level of the site, where alternating bands of bamboo and turf are interspersed with metasequoia trees and sculpture pieces. At the periphery of this space the ground plane is lifted with a berm that gives definition to this sculpture room before transitioning to the out of sight structures beyond. It is a simple yet effective gesture.

At a grander scale, the berm rings bounding the Circular Park at **Nishi Harima Science Garden City**, Japan (1993), have been constructed to give definition to an entire parkland, providing an armature for building a sense of place out of what had previously been a denuded landscape. This site envelops a roadway intersection, with the circular landform built at a scale reminiscent of a river embankment. The three wedge-shaped parcels contain offset arcs of paving, poplar trees, and hedges, along with parking areas for visitors to this remote location. Viewed from above, the park's form echoes the supercollider built just up the road that instigated development of this area in the first place.

Landforms define space within a once denuded landscape at **Nishi Harima Science Garden City**, 1993.
[Photographed during and soon after construction]

In experience the Circular Park feels carved out of the surrounding landscape rather than overlaid on what had been stripped terrain, a credit to the park's design and implementation. Reaching this site after climbing connected wooded and winding roads, the park also creates an unmistakable marker for the entrance to this new town. Even so, the impact is really most effective at a vehicular scale when approaching the intersection by bus or private car. Lacking a clear program or set of active uses, the park functions as a kind of spatial folly as part of the larger masterplan.

A grass-planted berm helps transform this functional space into an inviting outdoor room at **GSB Knight Management Center**, Stanford University, Palo Alto, 2011. [Photographed 7 years after construction]

A similar form of arced berm is found tucked away at the edge of **GSB Knight Management Center**, Stanford University (2011), only here reduced down to the pedestrian scale. Home to Stanford's business school, the center includes a collection of mid-sized buildings arranged to create a sequence of tony courtyards that make the most of Palo Alto's inviting climate. A linear band of plaza spaces offers spill-out areas for users of the adjacent buildings, with a mix of stone paving, tree shade, and comfortable seating. This axis terminates not at a grand structure but rather at an open area that is scaled to allow for turning emergency vehicles and yet is detailed to function as an additional gathering area. The berm built up at this outer edge helps contain the energy of the space and lessens the sense of exposure, meaning those enjoying the cafe seating feel like they are within a green room rather than open platform. An additional benefit of this landform is that it provides a smooth transition in grade to the boundary edge of the site that is at a lower elevation. Trees planted on the back side of the berm help diffuse this transition.

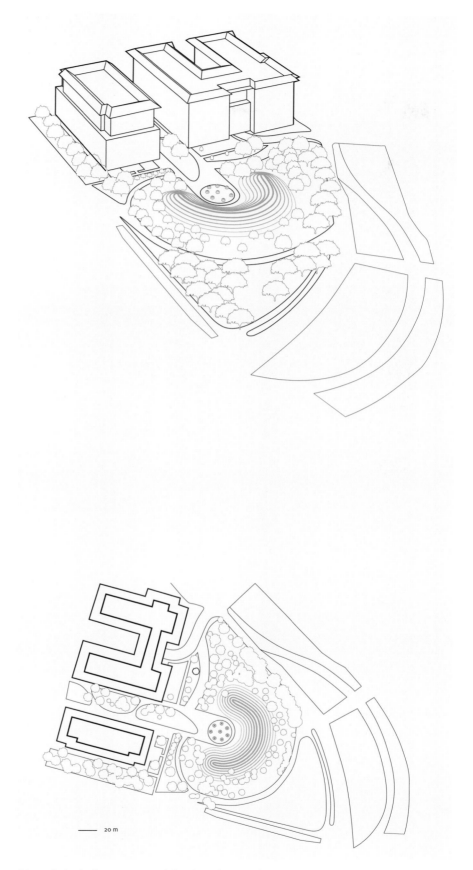

20 m

This semi-circular berm creates a defined terminus to a linear sequence of plazas that also accommodates emergency vehicles.

Weyerhaeuser Headquarters, Federal Way, Washington, 1972. [Photographed 46 years after construction]

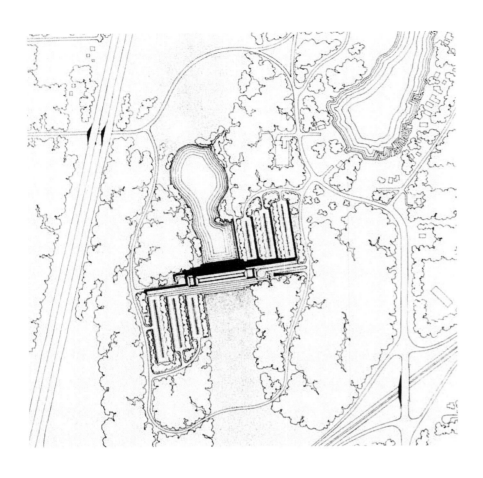

SITING AS STATEMENT

Choices made in positioning buildings within a site have lasting impacts on the experience and performance of those structures. For sites that are large relative to the necessary building footprint to be accommodated, these decisions will also impact circulation patterns, configuration of open spaces, and the overall relationship between the building and its surrounding landscape. Within Walker's portfolio we find some striking examples for how siting decisions made a statement about the project and its use.

The **Weyerhaeuser Headquarters** (1972) boldly expresses the potential for topography to choreograph the experience of a site, in this case a sequence sublime in character. Located outside of Tacoma, Washington, the project was built during an era when the corporate campus was ascendant[4] and organizations like this forestry company were looking to build on greenfield sites. The SOM-designed building lies perpendicular to two existing ridges, spanning a gentle valley in a way evocative of a dam. The structure is at once both in contrast to and integrated with the surrounding landscape.

Integrating landscape and architecture at **Weyerhaeuser Headquarters**, 1972. [Photographed 46 years after construction]

For this very much automobile-dominated context, the site is approached through a descending hierarchy of roads, exiting a nearby freeway before approaching the entrance via a perimeter road that orbits the main structure. This approach sequence gives visitors a full view of the site that then disappears as they move through forested areas, only to have the building appear again as one reaches the parking area—it is a cinematic procession that heightens the sense of arrival. The parking lots are built as terraces descending the valley wall and aligned to each connect with different levels of the main building.

Experiencing this approach sequence and exploring the grounds of the Weyerhaeuser site nearly a half century after its creation is still profound, with the site feeling like a well-considered composition in a way incredibly rare at this scale. This success is due first and foremost to the project's use of topography and integration into the site planning process. As a typology this project is out of step with current interests (the original owner has itself moved to downtown Seattle), but as a work of landscape design it still offers lessons ready to be learned and reinterpreted by a new generation.

Incorporating a reading of existing topography is especially valuable for site plans with major structures, but this responsiveness can also have an important role for landscape-driven schemes where buildings play a secondary role. When major landscape features are present, designers can make conscious decisions regarding how to connect with these natural systems and alternatively create newly built areas that stand in clear contrast. This calculated mixing of embrace and divergence is evident in some of the larger scaled projects produced through Walker's office.

The **Newport Beach Civic Center and Park** (2013) spans two linear parcels that are linked by a pedestrian bridge and descend toward the coast of this Southern California city. The project makes maximum use of a long-ignored fragment of land, bordered on one side by a major road and on the other by the Fashion Island Center whose masterplan was led by Walker in the 1960s. The park's grading scheme amplifies topographic variation in a way that supports specific program activities often positioned on the high ground while also guiding stormwater runoff, terminating with constructed wetlands at the site's low point. Preexisting wetlands provide a fulcrum around which the site's layout revolves, with gently sloped pathways encircling and then dramatically striking through these features. This tight relationship between topographic change, circulation, and major natural (or at least *naturalized*) features produces an experience of expansion and compression that defines the character of this landscape.

An additional layer integrated into the site's composition is a set of more intentionally manicured landscape spaces that stand in stark contrast to the restoration areas. Sauntering through the park (for really that is the most rewarding activity of this site) on a sunny afternoon, the juxtaposition of these different landscape types is more complementary than conflicting, with one framing the other in a flattering manner. This kind of scheme runs the risk of trying to do too much, but in this place and with this level of execution the project encourages ambition in continuing to evolve our expectations for built landscapes.

Contrasting conditions enrich the experience of landscape at **Newport Beach Civic Center and Park**, Newport Beach, California, 2013. [Photographed 5 years after construction]

Manipulating the ground plane is foundational to the shaping of outdoor spaces, and through a reading of Walker's extensive portfolio we can identify a number of takeaway lessons:

— Topographic manipulations can have an especially lasting impact on a site, as evidenced through some of his earliest works dating back more than a half century.

— Grading is most effective—or at least most *grounded*—when multipurpose and relating to additional systems such as circulation, drainage, and specific program needs.

— With increased knowledge and dexterity the designer might become self-conscious about associations of the designed landforms (i.e. pastoral versus minimal, etc.), resulting in even more intentional decision making.

— And finally, while site engineering directly benefits from increased computational capabilities, this does not necessarily lead to improved projects—some of Walker's most powerful landscapes were manually graded and executed by small teams.

1 From Walker's introduction to *Grading for Landscape Architects and Architects* (Basel; Boston: Birkhäuser, 2008), 8.

2 Stanford University, where Walker would in time design many landscapes is just downhill, and on from there technology campuses for the likes of Google and Facebook.

3 Peter Walker, *Invisible Gardens* (Berkeley, CA: Spacemaker Press, 1997), 123.

4 For more on corporate campuses see Louise Mozingo, *Pastoral Capitalism: A History of Suburban Corporate Landscapes* (Cambridge, MA: MIT Press, 2011).

IBM Silicon Valley Lab, 1977. [Photographed 42 years after construction]

IV

ON-STRUCTURE: FABRICATING ELEVATED LANDSCAPES

On-structure landscapes are inherently artificial, built above ground and disconnected from natural systems in a way that makes them perpetually vulnerable. These elevated green spaces are even more challenging *and* expensive to construct than other landscapes and yet they are becoming only increasingly prevalent. What is driving this continued adoption? Two factors: further demands on urban land where undeveloped green sites are rare can make the investment in on-structure landscapes deemed worthwhile; and second, improved technical capability and design expertise that makes the process of building these spaces a better-known risk. On-structure landscapes are maturing as a project type.

For all the variety of contexts and approaches surrounding these landscapes, there is consistency in the core challenges being faced with their implementation. There is a need to balance the soil (growing medium) and water needs for the vegetation with investment in supporting structural capacity. Plantings often face unique environmental challenges including increased exposure when growing in these inhospitable conditions. And finally, there can be complexities in managing access, with the need to make these removed spaces inviting to users while also often providing high levels of control to operators.

Walker's portfolio presents an especially valuable resource to the inquisitive designer since on-structure landscapes featured early in his work and went on to be a recurring project condition throughout his career. The result is a range of projects to investigate that contain within them design ideas that have been tested across multiple decades and often pursued in a surprising, non-linear manner.

CAPPING THE PODIUM

Alcoa Plaza (1961, now One Maritime Plaza) was one of the first on-structure landscapes for Walker to be involved with, constructed in the 1960s as part of the Golden Gateway urban renewal project in San Francisco that also includes Sydney Walton Square (discussed in previous chapter). The landscape is built upon an above-ground parking structure that spans two blocks and has an office tower rising from its center, with a mirrored array of rectilinear plaza spaces linked by an inner circulation loop. The plan is highly rational, to the point of looking *staid* from above, but this representation belies a richness in experience.

Heavily planted areas anchor the four corners of the site, with a circular seatwall at the center of these squares to offer quiet seating areas that are buffered from the surrounding urban environment. Closer to the main tower, more formal garden rooms are outlined by inconspicuous fencing, with perched pads of turf edged by terrace steps. These spaces all host works of sculpture. Meanwhile, larger expanses immediately adjacent to the tower provide visual relief, on one side presenting a simple plane of grass and on the other, a welcoming fountain.

Visiting the site more than a half century after its completion, it is striking to see just how substantial and uncompromising these spaces are and how well they have endured so many years of weathering and use. Some might see the landscape as "dated," and indeed there are telltale site elements like globed light fixtures that place this design in a different era. But on the whole the site is in distinctly fine physical condition, with the concrete walls exhibiting little to no cracking, finishes like the exposed aggregate unblemished, and even the timber used to provide some thermal comfort atop these walls still in place (albeit with many thick layers of paint). This endurance points to the value of

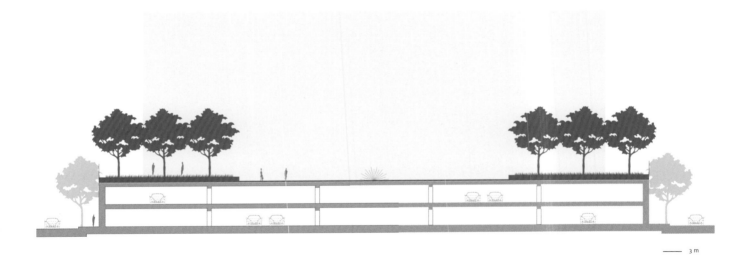

——— 3 m

Rudimentary in approach relative to more complex contemporary works, this garden built upon a parking structure still offers a richness of experience many decades after its construction.

approaching an on-structure project as a fully intensive endeavor, one in which the significant demands brought by heavy plantings, necessary waterproofing, ongoing maintenance, and "live load" activity of visitors is embraced. That is to say, there are no shortcuts, no ways of sneaking in active uses of the rooftop landscape. This role of the site is valued from the outset and planned for—designed for—accordingly.

At the same time, the no-nonsense approach that infuses the site's conceptual design all the way down to its fine detailing is well-executed and paired with relatively conservative use of materials: strong concrete walls with roughened finishes and chamfered edges; adequate volumes of soil for intended plantings; in rare uses of timber, a blockiness that brings integrity. Robust plantings go a long way in aiding this effect, softening the site's geometry while providing a layered buffer between this elevated landscape and its urban surroundings.

This is a privately owned green space, and its continued management and upkeep throughout the years must have aided its positive current condition. Something not readily apparent in published drawings but made clear when exploring the space in person is the important role that linked pedestrian bridges play in preventing this landscape to exude isolation in the way that can so often be experienced with other rooftop gardens. These linkages mean this private landscape is accessible without barriers to the public.

Simply adapting and repeating the main elements of Alcoa Plaza in different configurations could produce solidly successful built works, but this would not be pushing the medium (or discipline) forward. The most significant spatial limitation to this approach is that soils are built up above the rooftop level and these masses interrupt the flow of movement across the landscape. Overcoming these limitations of soil mass and walking surface was at the core of the next explored project, Saitama Plaza.

Alcoa Plaza in San Francisco, 1961, presents a highly legible example of an on-structure landscape where soil beds sit above the supportive podium and vary in thickness according to planting needs. [Photographed 57 years after construction]

ELEVATING THE FOREST

Saitama Plaza in Saitama, Japan (2000), is conceived as an elevated forest in the city, holding a dense grove of zelkova trees arrayed on an urban plaza that is perched three stories above surrounding streets and structures. As a lifted plane the space functions as a connector, providing pedestrian linkages to an assembly of rail transport, offices, shopping, and event spaces including a major stadium. What might otherwise have easily resulted in a lifeless space that was isolated and uninviting, here is instead made tranquil and alluring for individuals and small groups alike because of the carefully considered hardscape elements and, most importantly, the dramatic effect of the tree grove.

The great technical innovation offered by this plaza is a novel structural system that creates two extensive planes: one for the upper walking surface, and another 1.5 meters (5 feet) below to support a continuous soil layer that allows for trees to grow without the root constraints imposed by more typical tree wells or planters. This system maximizes growth potential of trees while also offering a continuous flat surface for pedestrians.

The "Forest in the Sky" concept of **Saitama Plaza**, Saitama, 2000, becomes apparent when approaching the space from an adjacent transit station. [Photographed 15 years after construction]

Among its many different roles, Saitama Plaza functions as a transit hub and as such there are varied means of arrival. To the southwest, an elevated train deposits travelers at an outdoor platform at effectively the same height as the plaza itself, visible from the distance as a "forest in the sky" just a short walk away. Even more transit riders will arrive via the station embedded within the east side of the site, where elevators release visitors at a walkway that gently descends into the main park space. Finally, for those relatively few arriving by private vehicle (most commonly on days of special events), stepped ramps link public sidewalks to the upper platform, with an extensive fountain running in parallel to mark the transition.

No matter what route is followed before entering the plaza, the forest of zelkovas has the appearance of a mass from a distance and entering through any of its sides offers a similar sense of arrival. Simple painted metal canopy structures mark out efficient walking routes for traversing the space while offering some protection from rain or sun.

The flatness of the plaza is notable when experienced in person, not just in the way it accommodates the movement of visitors but also for how it allows largely uninterrupted views across the space. With the masses of soil hidden beneath the walking surface, a greater emphasis is placed on the zelkova trees themselves. These trees, with their near-complete canopy coverage, present a degree of vitality and maturity rarely encountered with on-structure plantings. Experiencing this space during different modes of use—streams of workers during weekday morning commute, clusters of friends gathered in the evening hours, special events being hosted on the weekend—it becomes clear that the technical innovation of this project has resulted in a humane space.

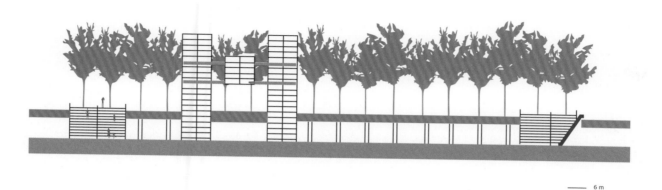

—— 6 m

This unique structural system supports an extensive soil bed that accommodates tree root growth while the upper plane creates a perfectly flat walking surface.

An original design model illustrates the linked layers of **Saitama Plaza**.

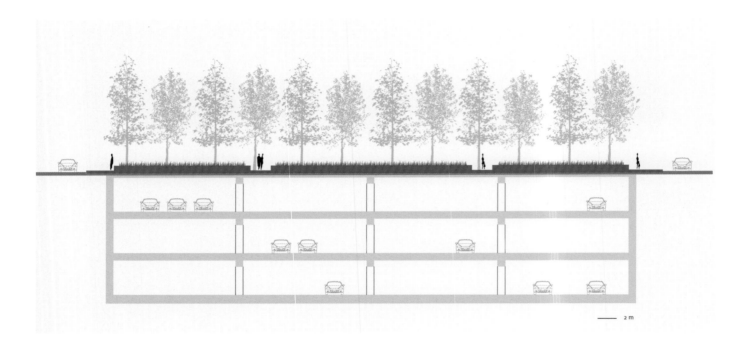

2 m

Back in California, another on-structure landscape completed just a few years after Saitama Plaza shows a dramatically different approach to flatness for an on-structure landscape, this time in the context of a university campus.

EXTENDING THE GROUND PLANE

Located within the Stanford University campus, the **Medical Center Parking Structure 4** (2003) landscape is embedded within an extensive medical complex. Not readily apparent in project imagery is the physical positioning of this landscape, which is indeed on-structure but effectively flush with the surrounding sidewalk and streets. The project plays with expectations of *terra firma* to the point that casual passersby would likely not guess this is in fact a rooftop landscape. The design mitigates negative impacts of an underground parking garage by allowing surrounding circulation to flow into this space.

The scheme works toward maximizing a sense of landscape while minimizing structural requirements. Plantings are largely limited to native grasses that while limited in height still together form a dynamic, inviting plane. In this way the landscape can be recognized as an extension of Walker's experimental Cambridge rooftops from twenty years earlier, only here there are fragments of grasses set within bench-height wood frames rather than silhouettes of metal tubing to evoke a field.

Despite being award-winning, this in the end is not an especially *exciting* project—it is a landscape that performs a particular role and as a case study illustrates how a relatively simple scheme can still play a valuable part within the greater composition of a university campus.

Medical Center Parking Structure 4, Stanford University, 2003. With the parking garage embedded below ground, this thin and relatively lightweight rooftop garden has the feeling of simply being a continuation of the surrounding landscape.
[Photographed 15 years after construction]

SPANNING CONNECTIONS

At the other end of California's Silicon Valley, tucked into hills past the southern reaches of San Jose, **IBM Santa Teresa** (1977, now **IBM Silicon Valley Lab**) presents an instructive example for how an on-structure landscape can play a central role in a campus masterplan, informing the arrangement of building massing to better support pedestrian linkages.

This is one of the earliest facilities built specifically to house teams of computer programmers. Eight four-story buildings containing offices for up to two thousand employees are clustered around what at ground level is a massive space for computer servers but above which is an open air plaza. An overall grid defined by the cruciform buildings is extended outdoors and bisected by pedestrian pathways. The resultant wedges of landscape are treated with a mix of turf, hardscape, or more significant planting areas.

As a form of crossroads for employees moving from one building to another, the plaza functions as an important social space that is urban in scale despite its rural context. The sense of connectivity is amplified by reflective glazing that makes up most of the plaza's perimeter, in addition to the inclusion of dining facilities that spill out with cafe seating. This restrained—almost austere—scheme is different than what would be built today, but maintains an exceptional level of integrity and clearly is maintained with great care.

For a showcase example of how different systems can be integrated at an especially ambitious scale, we will turn to Singapore.

The central courtyard at **IBM Santa Teresa** (now **IBM Silicon Valley Lab**), 1977, is built above central computer facilities and accommodates cross-building connections within this early high tech campus. [Photographed 42 years after construction]

The towering silhouette of **Marina Bay Sands Integrated Resort** has become iconic, visible throughout the Singapore cityscape.

Ways in which on-structure landscapes can become "the new normal" are on display at **Marina Bay Sands Integrated Resort** (2011), a massive and multi-use development in the tropical city-state of Singapore. Made up of a towering hotel, shopping mall, casino, convention center, and fully integrated underground mass transit system, this complex is draped in more than 12 hectares (30 acres) of landscape, the vast majority of which is on-structure. Key elements of this landscape include a 1.5 kilometer (approximately 1 mile) long waterfront promenade, podium deck, rooftop terraces, planted landscape bridges, and even a 57-story high roof garden paired with an extended infinity edge pool. It is a world-class development in scope and implementation.

At such a massive scale and with such a diversity of conditions, Marina Bay Sands makes use of many different on-structure strategies. Where feasible, clustered trees share planting areas and build from some of the lessons learned at Saitama Plaza with contiguous soils. Where structural conditions are more constrained, greater attention is given to the role of hardscape, with seating and shading structures making the environment more inviting. The relatively narrow rooftop terrace above the shopping mall is of particular interest, with lightweight wood decking transitioning to seating terraces in a way that allows small groups to gather while accommodating through traffic.

Planted pedestrian bridges span the center of the site, functioning as appealing linkages between the main parcels of the complex even if most visitors will likely elect for using the air-conditioned passageways running underground. These bridges, striking an elegant green line with their allées of trees when viewed from a distance, on closer inspection show evidence of trying to do too much within given constraints (physical and otherwise). Here the proportions feel amiss, with not enough space for the main pedestrian walkway nor the tree planters, even as utility equipment sits above the surface. Tree grate and other paving details have become jumbled to the point of warning tape being applied as a safety precaution against tripping. These imperfections are not enough to make the bridges unsuccessful—these still are well-used and tremendously expensive slices of landscape—but these misses stand out because they are so rarely found in the portfolio of works produced by Walker's office.

Meanwhile, the surfboard-like roof gardens topping the three hotel towers have become not only a social media-friendly backdrop for throngs of visitors, but a dramatic icon visible throughout the city. This linear platform holds an infinity edge pool along the inner edge of its gentle curve, from which arced bands of sunbathing terraces, a central walkway, and more sheltered leisure spaces radiate. Restaurants function as nodes distributed across the length of the space, terminating in a large viewing deck.

The massive **Marina Bay Sands Integrated Resort** in Singapore, 2011, contains some 12 hectares (30 acres) of landscape, the vast majority of which is on-structure. [Photographed 7 years after construction]

The experience of these rooftop gardens is more confined than might be anticipated, a result of so much program being packed into this space combined with the almost overwhelming level of popularity with guests. Even so, it is highly efficient and the kind of space that developers typically only dream of creating.

Marina Bay Sands Integrated Resort is a technical achievement and dramatic commercial success empowered by the allure of inviting on-structure landscapes, and yet in *experiencing* the site we can feel that something has been lost. Despite the thoughtfulness of the design and impressive engineering feats, these landscape spaces feel disconnected from their context. In this way the site represents the kind of spaceship-like mega projects that are only growing in prevalence within commercial capitals across the globe, and questions must be asked by designers and hopefully also their client groups as to what is ultimately desired. This form of hybrid landscape/architecture now made possible by innovations in design and engineering surely has a role to play in evolving forms of urbanism, but the value of legibility of place and rootedness in landscape ought not be forgotten.

Beyond connectedness, this project also reveals some of the strains in maintaining high levels of precision across such an extensive and technically demanding site, such as flatwork damage as a result of subsidence. Some of these issues may be due to installation flaws, but there also are signs of time pressure, for example where fine grading requirements were not aligned with paving material dimensions. By comparison, we could visit an on-structure site like Alcoa Plaza that shows less sign of wear despite being a half century older. These are two very different projects, to be sure, but the differing pace of change revealed through an examination of their spaces points to ongoing challenges of implementing high-quality works at both great scale and speed.

WRAPPING THE VOLUME

The on-structure setting for landscape at **Barangaroo Reserve** in Sydney (2014) is a surprise, where a sequence of park spaces that most would assume to be built on a natural landform in fact largely sits atop a voluminous cultural center and public garage. The building is almost entirely hidden from view, save for entry points at either end of the site that, if anything, suggest an underground space has been *carved out* rather than built up. With the overarching project goal (set before the designers were involved) of "recreating" the historic headlands that had long ago been removed from this site, this landform was created by building an extensive structure that was then capped and wrapped with vegetation.

A closer investigation of how this building volume transitions to the surrounding edges of the site is revealing. The most gradual transition is on the northern edge of the park, where near sea level, pathways link to grassy slopes that climb to the summit of the site. These constructed hillsides offer spacious areas for visitors to recline and enjoy views of the harbor.

On the western edge, the transition from high point to shoreline is compressed within a sequence of heavily planted terraces. Switchback paths within these terraces give visitors access to this area and make it a valuable garden in its own right. Meanwhile, the park's main volume is most abruptly terminated to the south. This edge serves the pragmatic purpose of providing the most unobstructed entry to the enclosed event space while also creating efficient stair and elevator access to the park summit.

The extensive green spaces at **Barangaroo Reserve** in Sydney, 2014, are largely built atop a cultural center, with varying edge conditions for descending to the surrounding shoreline.
[Photographed 4 years after construction]

Barangaroo Reserve offers lessons for how a landscape-driven project can still necessarily be responsive to artificial conditions, using design to tightly link structural demands with desired landscape experiences.

With continued urbanization across the globe, there will be only increased demands for constructing green spaces in contexts not inherently suitable for landscape. Reviewing Walker's portfolio we can recognize a clear evolution of on-structure landscapes across the past six decades, with increases in scale and sophistication aligned with an expanded diversity of contexts in which these environments have been created. The takeaway lessons are both technical and conceptual, including:

— Synthesizing concerns regarding structural capacity, soil volumes, and means of access is central to the design of any on-structure landscape.

— The separation of walking and drainage surfaces associated with these sites offers distinct opportunities for flatness not possible with terrestrial landscapes that must be graded to capture runoff.

— In making use of on-structure systems, there are conceptual decisions to be made in how much this artificiality is expressed versus hidden.

— Designers must take on the challenge of addressing both the possibilities but also potentially negative consequences of making use of on-structure landscapes in project sites that are increasingly disconnected from their immediate context.

Center for Advanced Science and Technology, Hyogo Prefecture, 1993. [Photographed 21 years after construction]

V

CRAFT IN DETAIL: SOPHISTICATED ENDURANCE

Craft—the skilled implementation of design—can be difficult to achieve in any discipline, but landscape architecture has its own challenging forces that must be addressed. Most especially, built landscapes are by definition exposed to the elements of climate: sun and its corresponding ultraviolet light radiation; air pollution; rain and other forms of humidity; plus ice and snow with their freeze and thaw conditions. And these are just the environmental challenges—there also are the impacts of activities, both in the programmatic usage of the landscapes but also ongoing maintenance that can bring additional forms of wear. In short, any work of design must embody a degree of robustness if it is going to survive over time and if it need not be robust, perhaps it is best thought of as art. Landscapes are exposed to especially varied vectors of impact.

Even as craft in the built landscape is difficult to achieve it also, paradoxically, can easily be taken for granted. This can be due to a mix of high and low expectations—*high* when viewing the landscape in comparison to far more easily controlled interior environments, and *low* when relating these spaces more generally to the greater built environment.

With all these challenges, what then might be considered the value of craft in landscape architecture? At the very least it can lead to improved longevity of the project, both in maintaining its supported activities but also its value as a community asset. Well-built landscapes have a better chance of being long-lasting landscapes.

Landscapes embodying ideals of craft can also be a sign, a form of test even, for the commitment of a client. Achieving a quality of craft cannot be an afterthought, and the requirements that must necessarily be addressed early in a project's process can allow a design team to learn more about the client's understanding and investment for the built landscape.

Craft also allows a landscape to rise above the everyday, to stand out. This need not be simply a vanity exercise—these parks and plazas can offer an experience unique compared to what else is available in the community, spaces that are likely to have been developed with minimal input from designers. Respect for material qualities, thorough detailing, and skilled implementation can elevate built landscapes from the mundane and transitory to reach toward the enduring and sublime.

In the end, craft-driven landscapes need not (and indeed *can* not) be the default approach taken with most professional works, but it ought to be recognized as an option. And it was this option, this subset of the built environment that Walker pursued most especially in the latter half of his career and from which we have a great many lessons to learn.

ELEMENTAL MATERIALITY

The expression of innate material qualities can be powerfully paired with craft of implementation, where the richness of the materials themselves help allow for a more subdued or "non-striving" (as Richard Haag might describe it) composition. An illustrative example of this is found within the central courtyard at the **Center for Advanced Science and Technology** (also referred to as **CAST**, 1993) located in Hyogo Prefecture, Japan. This garden—described in the anecdote at the beginning of this book—is meant to be viewed more than inhabited, with the space visible through windows and (typically locked) sliding doors surrounding the space. Turf, stone, and wood are used as solid, integral elements within this walled plane, and as visitors move through the adjacent sequence of linked hallways they are presented with shifting views of the courtyard. In this way the internal landscape functions as a core around which the buildings of the complex orbit.

A fair question to ask when only presented with images of the courtyard would be, is this really enough? Conceptually, sure, there is an elegance to the restraint of the scheme, but is that still satisfying in person? In a word: yes. This research facility is positioned in a remote location with forested hills,[1] and before reaching the courtyard an individual will have travelled through a range of broader landscapes. This prefacing helps prepare the visitor to appreciate a distilled referencing of that greater environment, emphasizing a sense of arrival.

Immediately adjacent to CAST and within the Harima Science Garden City masterplan, a public space envelopes the intersection of two arterial roads to create the multi-quadrant Circular Park (1994). A mysterious hemisphere of stone rises from one of these wedges of park, clearly visible from the road but also inviting—and *rewarding*—closer inspection. The object exhibits a play of scale: from a distance it dominates this space to celebrate the nearby supercollider, albeit with a roughened, almost reptilian surface; from up close, the immense mass of each individual stone is made more evident, and yet the sphere is inviting to be touched or even climbed. The stones are placed without mortar, allowing for an inner light source to shine through these gaps at night. The stone hemisphere at **Harima Science Garden City** illustrates how an assemblage of individual material elements can be combined to form an iconic object in the landscape.

Material expression in the central courtyard at the **Center for Advanced Science and Technology**, Hyogo Prefecture, 1993, complete with indoor-outdoor connections that beckon the eye if not always the feet.
[Photographed 21 years after construction]

Stone boulders are interlocked without mortar to create an iconic element at **Harima Science Garden City**, 1994.
[Photographed 18 years after construction]

Expressions of stone at **IBM Japan Makuhari Building**, Makuhari, Chiba Prefecture, 1991. [Photographed 23 years after construction]

The landscape fronting **IBM Japan Makuhari Building** in Makuhari, Chiba Prefecture, Japan (1991) can be appreciated from many viewpoints, and as such the site presents an especially fitting setting for an elemental use of materials that is paired with refined implementation. Set within a suburban district reachable by commuter train from Tokyo, most visitors arrive to this facility on foot via a network of surrounding sidewalks and bridgeways. This elevated point of entry highlights the courtyard design and allows for a greater range of perspectives into the garden. Notably, there is almost no use of wood, save for a trellis structure only accessible from the building's interior. Instead the site is largely a play on stone and vegetation, where design gestures like the rows of hedges[2] transition to slate-clad (and with that green-hued) seatwalls on the upper walkways. IBM Japan Makuhari Building offers an example of a project where the landscape, despite having relatively minimal program requirements and therefore limited practical importance, in the end plays an outsized role in defining the character of this site relative to the elegant but nondescript main building.

Paving detailed to express hierarchy of materials at **IBM Solana** in Texas, 1989, **Sony Center** in Berlin, 2000, and **Toyota Municipal Museum of Art**, 1995. [Photographed 29, 3 and 23 years after construction, respectively]

RESPECTING EDGES, EMBRACING HIERARCHY

Craft can mean a respect for the cohesion of a material that leads to decision making concerning the hierarchy of elements in hardscape detailing. A clear, if simple, example of this is found at the pathway edge near the main water feature at **IBM Solana** in Texas (1989). Small boulders of the same stone as used along the water's edge are scattered into the adjacent planted area, giving a sense of this being a zone or environment rather than just an assembly of elements. The effect is emphasized by having some of these boulders overlap—*interrupt*—the bent steel edging of the decomposed granite paving. The hierarchy of elements can also be seen in paving details at the **Sony Center** in Berlin, where stainless-steel panels are kept whole and cobblestone paving becomes secondary. Meanwhile, within the upper terrace at **Toyota Municipal Museum of Art** (1995), a pathway of cut slate stone maintains its sequence of whole units while projecting into the more amorphous asphalt paving.

IBM Solana in Texas, 1989.

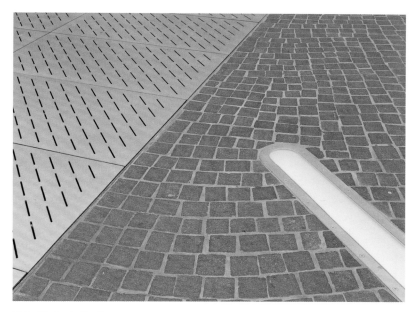

Sony Center in Berlin, 2000.

Paving edging interfacing materials and delineating space at **Toyota Municipal Museum of Art**, Toyota City, 1995.
[Photographed 23 years after construction]

Continuing with Toyota, paving edges throughout the site present a rich array of examples that are infused with craftsmanship and embody landscape sensibilities. Fist-sized granite stone is used both to define pathways and also distinguish planting areas, such as between turf and bamboo groundcover. The light color of this edging material contrasts the green vegetation and dark grey asphalt to create a bright frame that acts as a graphic overlay across the garden. Meanwhile, the hardness of the material allows it to be hand-chiseled and installed with a mix of precision and looseness that is especially appealing.

At the lower level of the site, meanwhile, this mindful mix of materials is used to transform what ordinarily would be a pragmatic vehicular drop-off and parking area to instead function as an inviting garden terrace. Forgoing painted stripes for a grid of rounded granite domes that are embedded into the asphalt amplifies the effect. Visiting the site while it is actively being used to host a regularly occurring weekend market makes a compelling case for how the creation of well-crafted landscapes creates opportunities for a range of future uses.

MUNDANE ELEMENTS, FINELY IMPLEMENTED

Hardscape detailing at the Newport Beach Civic Center and Park (2013) presents more recent examples of craft that make use of common materials. At the upper area of this stepped, multi-parcel site an off-leash dog park is bounded by a distinctly sculptural fence. Using a running sequence of embedded vertical elements is an elegant form of fencing that can be found in many high-end institutional sites, including other Walker projects. As commonly detailed these site elements are relatively expensive, with the vertical (and often rectangular) metal bars welded into a base plate that is mechanically attached to an underground footing. Here at Newport Beach, however, those vertical elements are made of lower grade steel formed as less expensive round tubing, protected with a powder-coated finish and embedded directly into a concrete base. The footing is brought above ground, functioning as a low wall that looks perfectly in place amongst the surrounding poured concrete flatwork.

This fence illustrates how techniques developed in higher-end landscapes can be adapted for use with more common materials to develop site elements that rise above the conventional.

Powder-coated steel poles embedded in a cast concrete base form an elegant and relatively economical fence at the **Newport Beach Civic Center and Park**, California, 2013. [Photographed 5 years after construction]

The **Library Walk**, University of California at San Diego (1995) relies on a restrained palette of modular units that are carefully applied and repeated at an extensive scale to create a unifying linear open space. This is another example of what could be perceived as almost a mismatch between the mundane quality of materials and exquisite implementation. Under other circumstances it could be cut stone rather than concrete unit pavers used for the walking surface, and perhaps stainless-steel boxes[3] rather than monolithic precast concrete blocks used for the bench-like structures metering out this axis. But with the pavers installed atop properly compacted grade and within well-anchored edging that minimizes lateral movement, these more economical elements still produce a crisp, refined effect. At the same time, the larger concrete blocks are sized so that their width equals a full multiple of pavers, allowing for the edge of these elements to align with paving joints. It is a subtle consideration, but when repeated and successfully implemented across such an expanse this detailing infuses the landscape with a high level of integrity.

The Library Walk at University of California at San Diego offers a reminder that there are important distinctions between simple and easy, and that through mindful detailing and proper installation a landscape space can rise above the class of its materials.

Clean alignment of preexisting and added paving elements at **Library Walk**, University of California at San Diego, 1995.
[Photographed 22 years after construction]

Detailing of paving edge allows for shadow between walkway and adjacent turf, emphasizing flatness at
Burnett Park, Fort Worth, 1983. [Photographed 35 years after construction]

ELEVATING THE CIVIC

A high level of craftsmanship in the public realm can elevate parks and
other civic works to be something that stands out amongst their peers.
Relative to privately funded and maintained landscapes, public projects
will often have more limited funding and it therefore is necessary to be
especially selective in how and where craft is applied.

One example can be found at **Burnett Park** in Fort Worth (completed
in 1983, with revisions in 2010), where paving detailing and in particular
the transition from the main walkways to the site's signature panels of
turf give a crispness to the landscape that is rare in any built landscape,
let alone a public park. As a hardscape element made up of cut stone and
a concrete base, this critical element is also highly robust and is still in
perfect condition decades after original construction. This site element—
key to producing the desired effect at the heart of the project's conceptual
design—has proven to be a worthwhile element to have directed so much
attention and resources toward.

At **Jamison Square** in Portland, Oregon (2006), it is the use of a generally weaker (or at least more vulnerable) material that offers an experience
distinctive for the public realm. Here a continuous wood-plank board-

walk [4] makes up the park's eastern edge, offering a pleasant surprise to pedestrians making their way through the surrounding city blocks and otherwise expecting a generic concrete sidewalk. The use of wood in this setting is unique in part just because of its higher costs, but also because poor detailing or installation could make this public space a liability. In this case, however, craft in implementation has allowed the feature to age gracefully even after a full decade of public use while also being detailed to meet the appropriately stringent needs of accessibility and safety.

This linear space illustrates how an embrace of craft can empower the use of unconventional (and generally less-forgiving) materials in a way that offers unique experiences within a public landscape.

Craft of detailing and implementation allows for unique experience of wood-plank boardwalk as public sidewalk at **Jamison Square**, Portland, Oregon, 2006. [Photographed 11 years after construction]

Within an institutional setting such as a university campus, high-quality and integral materials that are implemented with relatively simple detailing can result in landscape elements that are robust while also distinguished from their immediate context. This is the case with the linear landscape sequence constructed as part of the campus landscape enhancement for the **University of Texas** at Dallas (2011). Examples can be seen in some areas of paving, where precast concrete pavers in an elongated form help give a textured feeling to a plaza that is more visually rich than conventional concrete paving. The narrowness of these pavers makes them susceptible to cracking under any kind of load, and therefore attention must be paid to specifying the (hidden) thickness in the z-dimension that makes each unit stronger and safely allows for a range of active uses. Being precast units, it is relatively easy to add integral coloring, and in this case the designers have specified a green blue that is slate-like in effect, and nicely complements the golden coloring of the stone seatwalls used in a terraced gathering space and also as independent linear objects demarcating space along the central axis of the campus. The spatial and functional needs for all of these areas could have been accomplished with more common materials, but the result would have been of lesser value. As it stands, especially in the context of a sprawling campus that is a public facility, this craft-infused landscape gives the community something to be proud of while (almost surprisingly) helping allow for other more secondary spaces to remain simpler in their design and implementation. For a facility such as this one, we can see the value in having resources somewhat concentrated and targeted toward the most important public spaces rather than diluted across the entirety of the campus.

Stone seatwalls and colored precast concrete pavers enhancing the campus of **University of Texas** at Dallas, 2011.
[Photographed 7 years after construction]

DEFINING THE PINNACLE

While limitations for implementing craft in the public realm can demand a high degree of judgment, what might be strategies for projects at the other end of the spectrum, i.e. with private works that are likely to have significant construction budgets and reliable long-term maintenance? Walker's intentional and largely successful pursuit of institutional clients since the 1980s has led to many opportunities for testing possibilities brought by such projects, the built results of which we can investigate and learn from.

Examples from Walker's portfolio include a collection of works within Novartis Campus,[5] headquarters for the Basel-based pharmaceutical giant. For some of these spaces, such as The Forum at Novartis Campus (2007), there is restraint and understatement in how this investment is employed. Mostly it is a matter of giving preference to high-quality natural materials such as stone over concrete, and even monolithic stone in lieu of stone cladding. The result is a landscape where a high level of craftmanship is the distinguishing characteristic. Moving through this space on a fine summer afternoon, the sensibilities of designer and (Swiss) client seem well aligned.

The effect and recognizable value of extra effort and investment is not always apparent, however, as suggested when examining the Sculpture Garden at Novartis Campus (2005) at the northern end of the Campus. This space cleverly integrates access control within a pedestrian underpass, allowing employees to enter the property (and technically cross the French-Swiss border) through an elegant sequence that culminates with a Richard Serra sculpture at campus entry. The stairs and walls in particular make use of craft in their concrete detailing, but the end result is somewhat underwhelming, cold even, and passersby are not likely to guess the level of investment that went into these spaces. Conversely, within the Nasher Sculpture Center in Dallas (2003), an institutional project where hardscape elements play a largely supporting role to vegetation, water, and of course the displayed artwork, craft still helps support greater integrity and functionality.

Monolithic stone paving with integrated drainage channel at its perimeter within The Forum at Novartis Campus, Basel, 2007. [Photographed 11 years after construction]

Concrete walls finely implemented but less clearly exhibiting their value within the **Sculpture Garden at Novartis Campus**.
[Photographed 13 years after construction]

Implementation of craft at **Nasher Sculpture Center**, Dallas, 2003, where even in a supporting role it infuses the site with greater integrity.
[Photographed 15 years after construction]

Determining when and where to employ craft in the built landscape requires critical consideration of context, uses and, ultimately, the hoped-for role that the landscape will play. As has been explored in the preceding projects, the intentional selection of materials and their careful implementation can on their own help elevate a project to something beyond the ordinary. Additionally, the mindfulness required by craft can result in improved weathering and greater endurance across the lifetime of a built landscape.

A benefit of our longitudinal study is that we can recognize the mix of consistency and variations throughout this one individual's career. With regards to craft, in his first phase of works (c1958–1978) we can identify a general humility, an acceptance of limitations in the use of landscape materials. Here there are sensibly scaled cast-in-place concrete walls with generous chamfers on their corners, but at the same time thoughtfulness is given to alignment of joint work and finishes. Where wood is used, such as for benches or in decking, the members are generously scaled to hold up well over time. In Walker's second major phase of practice (c1979–2015), the shift in focus from developer-driven to institution-led projects brought a general upgrading of materials: less concrete and more stone; less painted or galvanized steel and more stainless steel. In these works we are offered an iterative testing of how materials can be features in themselves, utilizing increased budgets to make more extensive use of natural materials. And finally, with the most recent designs being developed as the firm founded by Walker enters a new phase (c2016–present), we see a testing of the limits for how much this quality of craft can be scaled up to increasingly large and ambitious project sites, distributed across the globe.

The legacy of Walker's landscapes challenges us to sharpen our thinking on the role of craft while at the same time offering inspiring examples of technique. Ultimately, craft is about process rather than necessary expense.

1   Or more precisely, largely re-forested that had been made bare during intensive grading exercises and later planted again according to the masterplan developed by Walker's office.

2   Walker has described the concept as being inspired by the punch cards used in early mainframe computing.

3   For example, as built within Walker's landscape at the Sony Center in Berlin.

4   The boardwalk is a nod to this district's industrial past and in PWP's masterplan was originally intended to be a unifying element spanning multiple blocks.

5   Walker's firm led efforts on this campus masterplan, one of many examples where early involvement at the planning stage ensued in later site design opportunities.

IBM Japan Makuhari Building, Chiba Prefecture, 1991. [Photographed 23 years after construction]

VI

VEGETATION: ELEMENTAL DEFINITION OF SPACE

Vegetation and the use of plantings are inextricably linked to notions of landscape architecture. Second perhaps only to topography, this is the most essential landscape component to be employed in the shaping of outdoor environments.

Assessing the value of plantings can prove challenging, especially when considering the full range of project stages. On the one hand, when revisiting a site that has been in place for many years, it often is the vegetation that reveals itself to have had the most lasting influence. And yet within the early stages of the design process there always is the risk of the plantings being undervalued, in part due to challenges in representation and communication with decision makers but also potentially as a result of limited expertise within the design team itself. In between these early and late stages there are the efforts of procurement, fulfillment of necessary budgets (for these can be significant), and the installation procedures themselves. Despite all this effort, plantings will typically look their worst at a project's opening but ought to improve with time—that is if properly installed and serviced with ongoing maintenance. Landscape architects are not likely to be successful if their expertise is limited to only one of these stages. Vegetation is a finicky ally of design.

As living materials, plantings become part of their surrounding ecology whether or not this engagement guides a design. Decisions must be made regarding how performative a landscape is intended to be as relates to spatial (or even just visual) considerations. This is never fully an either/or situation, and the type of performance sought can take many forms. It can primarily be about limiting the impact of urban development on surrounding natural systems, for example, where stormwater detention is integrated with newly formed public space. Or the focus could primarily be about creation of new habitat, reducing urban heat island effects, carbon sequestration, defending against ocean storm surges, and more. This performance can take the form of productivity with urban farming or other food-related themes. There are no hard definitions for when these projects slip out of the purview of landscape architecture and into their own specializations (i.e. habitat ecology, stormwater engineering, or indeed ... farming), but when the sites are meant to engage with the public in some meaningful way and are not solely focused on ecological performance, it is in these cases that landscape architects will often have something to offer.

Some may approach an investigation of Walker's portfolio with a sense that plantings have not played a significant role in his efforts, an impression perhaps due to the times he has focused on especially minimalist works. A closer examination of what has been built reveals that vegetation has in fact been used in a rich variety of ways and with a clarity of purpose. Planting selections fulfill specific aims in a way that renders these uses as distinct design strategies.

An existing forest grove was preserved and augmented with additional plantings to create a lush perimeter at **Toyota Municipal Museum of Art**, Toyota City, 1995. [Photographed 17 years after construction]

CONTROLLING THE PERIPHERY

Perimeter planting can be important in controlling the design and experience of a site's landscape spaces. This planting that serves as a buffer to outside elements might already be in place, either naturally occurring or as a legacy from previous works. In other situations the perimeter trees can be an intentional addition, even if the end result has an informal appearance. At **Toyota Municipal Museum of Art** (1995) it is a mix of old and new that effectively establishes a green frame for that site's carefully crafted garden spaces. As a place created explicitly for experiencing sited sculpture works its surrounding band of densely planted trees creates a valuable backdrop analogous to museum walls, only here instead of the neutrality of white paint it is a dynamic patchwork of green foliage. When moving through the site's sequence of gardens the layered informality of the perimeter planting also provides a welcome counter to the more formal and geometric landscape layout. If the perimeter of the site was simply a wall of one form or another, the site's planting design would run the risk of being overly austere. As it is, however, these spaces have the feeling of refinement within a humane and inviting forest ring. Perimeter planting can create a complementary contrast to other landscape elements.

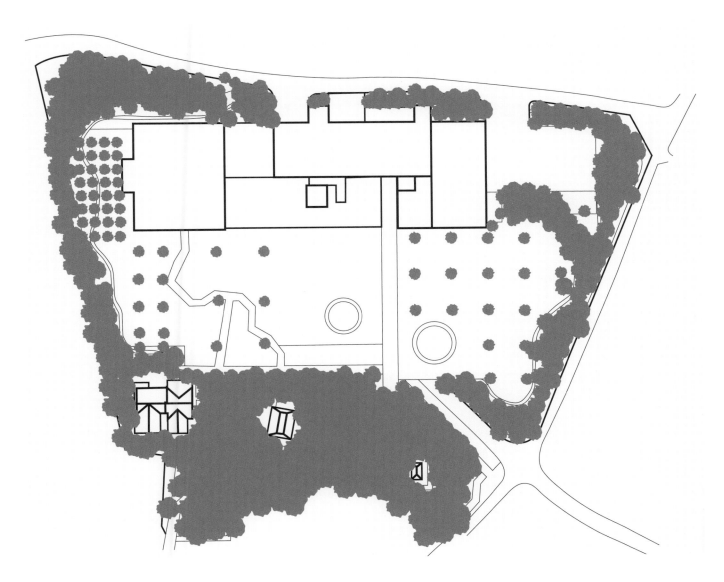

Urban constraints and the logistics of construction can make project sites something of a *tabula rasa*, where all existing vegetation has been erased. In these situations screening rows of tightly spaced trees can prove especially effective for marking a site's boundaries and controlling the experience from within the designed spaces. A dramatic example of this approach can be seen at the **Nasher Sculpture Center** in Dallas (2003) where broad-leaved and evergreen magnolia trees mark the project's property line. These trees create a striking (if not especially *inviting*) green wall when viewed from the public street while, even more importantly, establishing a tall green backdrop to the garden spaces within this tightly configured museum. The height of the trees mitigate the looming presence of surrounding building towers and as middle-ground elements when viewed from within make the sculpture garden feel more expansive. Perimeter planting can help a site feel larger.

A hybrid of these two approaches to perimeter planting can be seen at **Jamison Square** in Portland (2006). Mature street tree plantings on the western edge of this city block function similar to the dense magnolias at Nasher, only in this case limbed up to accommodate pedestrian movement beneath the branches. On the eastern edge of the site it is a more diffuse planting of honeylocust trees that functions as a welcoming entry rather than hard border to the site. The selection of a somewhat informal *robinia* pairs nicely with this area's paving materials, wood decking for the sidewalk, and decomposed granite for the adjacent sculpture rooms. This edge of Jamison Square illustrates how perimeter planting can function as a porous interface.

Perimeter planting at **Jamison Square**, Portland, 2006, provides a diffuse boundary that defines the interior space while welcoming visitors to enter this urban park. [Photographed 11 years after construction]

Magnolia trees at the perimeter of **Nasher Sculpture Center**, Dallas, 2003, are spaced to provide a nearly continuous screen that frames the site's sequence of outdoor rooms. [Photographed 15 years after construction]

Pin oak trees are sized and arranged to define a volume of space within **The Forum** at Novartis Campus, Basel, 2007.
[Photographed 11 years after construction]

DEFINING SPACE

Perimeter trees can be effective in defining space like the walls to a room, but what is even more evocative to the experience of landscape is the use of trees to define *volumes* of space. Two distinct approaches to the definition of volume can be found at the **Novartis Campus** in Basel. Within a courtyard referred to as **The Forum** (2007), more than thirty mature pin oak trees are arranged on a grid to form a bosque. Because of their height, the canopy formed by these trees is high enough for the free movement of visitors below the branches in addition to allowing for visual access across the space. With careful consideration of this species' growing habits, the individual trees are spaced to enable nearly complete canopy coverage. The ultimate effect is the creation of an outdoor room that is at once grand in scale while also inviting and intimate due to the relatively low "ceiling" of the branches.

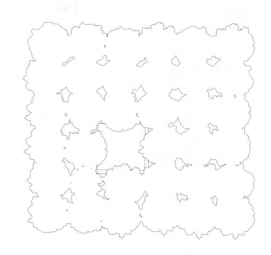

Quite a different, and in some ways even more provocative, approach to using trees to define space can be found just a short walk from The Forum, within what is known as the **Forum 1 Courtyard** (2004). The northern half of this space is dominated by an informal (but mind, not *random*) collection of Himalayan birch trees that with their svelte trunks create a mixed opacity screen. These vertical obstructions have the surprising effect of making the space in fact feel larger, since the confines of the courtyard are never fully visible. With moveable cafe furniture distributed amongst these pale tree trunks, small groups of employees can sit in conversation with a feeling of semi-privacy. The overhead canopy formed by the birches adds to the effect, screening direct views from colleagues that may be looking down from the surrounding offices. It is urban and urbane.

A ring of clipped European hornbeam at the southern edge of this modestly scaled courtyard offers yet another prototype for the use of trees in providing spatial definition, in this case creating a circular void that acts as an unmistakable counter to the adjacent mass of birches. One might expect such a well-defined "room" to feel like a place to be still, but in experience the visitor feels exposed in the space and it is far better suited to support formal walkways as are there now. It is easy to imagine a less experienced design team reversing this arrangement of mass and void but doing so would entirely miss the design goals of this landscape.

Plays on mass and void with tree plantings in the **Forum 1 Courtyard** at Novartis Campus, Basel, 2004.
[Photographed 14 years after construction]

Aerial view illustrating clustering of trees around the main water feature at
IBM Solana in Texas, 1989, soon after construction, along with more contemporary
view of these same trees in winter. [Photographed 29 years after construction]

COMPLEMENTING WATER

Tree selections, especially when used more individually rather than in large clusters, can be placed to make subtle thematic references. This is especially evident with plantings that are paired with water features.

One of the joys in developing a longitudinal study of a portfolio as extensive as Peter Walker's is the chance to see certain ideas iterated on in adapted form within projects that may be many decades apart. A case in point is the use of willow trees near water. Most species of willow are in fact water-loving, and in a more naturalized landscape where a pond is revealing a perched water table the existence of willow trees can point to the presence of wet soils. Within the constructed landscapes of urban environments, however, there almost certainly is a separation between the visible water features and soil systems, so why then might these trees still be planted?

In Walker's landscapes we can interpret a reasoning of subtle association, the kind that most visitors are not at all likely to explicitly notice but that targets some more deeply embedded understanding within us all. For a designer who has often challenged design expectations and been more than willing to objectify planting material in his landscapes, this almost conventional pairing of water feature and willows could be surprising, but it also is humane and inviting. The suggestion can be that even within an ambitious landscape design there is a need for give and take, with the calming effect brought by these paired elements in some ways buying opportunity for more unexpected elements in other areas of the design.

Three different uses of willows illustrate how a change in scale can impact the effect of these trees. At the signature water feature of IBM Solana (1989), the linear grove of willows creates a destination for employees who venture out from their offices and across the adjacent parking lot. Something not readily apparent in project photos but very much felt when visited in person is the sense of exposure in this landscape, due mostly to the lack of building mass in the immediate vicinity and the near complete flatness of the site itself. In this context the massing of trees creates a hospitable zone that would be far less effective if the willows were fewer in number and more sporadically spaced.

Half a world away from Texan IBM Solana and at the periphery of the Tokyo metropolitan region, **IBM Japan Makuhari Building** (1991) displays yet another examplary use of willow trees paired with water. In this case it is a trio of trees placed within a planter island that is surrounded by water. The light-colored gravel base of this platform provides contrast to the vibrant green of the willows and emphasizes their sculptural form. If fewer in number, say only one or two trees, these would look vulnerable while at the same time not providing enough mass to fully inhabit the space. At the same time, increasing the number of trees to form a more densely packed grove would detract from the trees' sculptural offering. This calibration of scale and number allows the trees to present a sense of ownership in occupying this courtyard space.

A variation on this approach to clustering willow trees can be found at the **Nasher Sculpture Center** (2003). Located in downtown Dallas, this project is an oasis in feeling but entirely urban in its efficiency. In fact the original design concept was significantly different from what is in place now, with sweeping terrain and prairie plantings in lieu of the tightly configured green rooms that were ultimately implemented. With the demands of siting many sculptures and accommodating large numbers of visitors, there is not the luxury of space to have either the large-scale clustering or more delicately positioned arrangement as expressed in the previous two examples. Instead, here at Nasher a double row of willows is embedded into a berm at the back edge of the site, complementing the rectilinear pond and working together with the terraced mass of the berm to create not just a backdrop to this space, but in fact a destination that rewards individuals who have traversed the whole of the garden to reach this terminus.

Stepped terraces of willows adjacent to water provide a perforated backdrop at **Nasher Sculpture Center**, Dallas, 2003. [Photographed 15 years after construction]

At the **Toyota Municipal Museum of Art** (1995) we can see an even more selective use of a willow paired with water, in this case as a single specimen planting. Compared to previously discussed examples, this site presents an interestingly different scale ratio between the two elements, with the main water feature especially extensive to the point where it has the appearance of a pond (although it is in fact built upon a rooftop of museum spaces below), and there being only a single willow tree. It might be expected that this lonely specimen appears overwhelmed by the scale of its surroundings, but in actuality its role is promoted, providing an accent, a piece of landscape jewelry that watches over the space. The tree's placement within a band of turf supports this effect, as does the mass of the surrounding forest that acts as a slightly removed backdrop to the specimen when viewed from the main path adjacent to the water feature.

Willow trees planted within the entry court at **IBM Japan Makuhari Building**, outside of Tokyo, 1991. [Photographed 23 years after construction]

An individual willow accents and looks over the main water feature at **Toyota Municipal Museum of Art**, 1995. [Photographed 17 years after construction]

Birch trees are positioned to complement the water feature at Jamison Square, Portland, 2006.
[Photographed 11 years after construction]

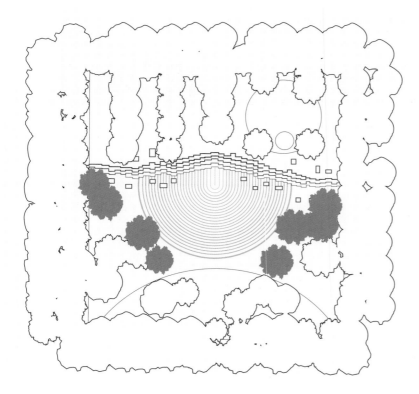

Meanwhile, at Jamison Square in Portland (2006) we find birches rather than willows complementing water. Although surrounded by paving and many steps away from the site's main water feature, these svelte trees with their flittering leaves and structural informality are notably (and perhaps even *surprisingly*) effective at infusing the space with a riparian atmosphere. Passing through the surrounding evergreen tree cover and descending the gently sloped basin toward the site's main gathering space, these birches stand in stark contrast to heighten the sense of arrival, of discovery.

A band of bamboo provides a dynamic backdrop to the entry courtyard at **IBM Japan Makuhari Building**, Chiba Prefecture, 1991, screening the building while also creating an atmospheric view back into this space.
[Photographed 23 years after construction]

AS SCREEN

Bamboo presents a special case when reviewing Walker's portfolio as the material is not commonly used in his landscapes, but when employed we can identify clear *intention* of its use. Even more than other types of plants, bamboo often has a particular role to play.

Or perhaps even two roles. At **IBM Japan Makuhari Building** near Tokyo (1991), a row of bamboo demarcates the northwestern edge of the site's main courtyard. Finely scaled and yet uniform in mass, this material creates a breeze-swaying green wall that is one of the few dynamic elements in this otherwise static garden space. From within the building, the bamboo filters views out onto the courtyard for employees dining within the canteen. The effect is highly atmospheric, also made possible by the floor to ceiling glazing. This project illustrates how bamboo can function as an opaque screen when viewed from a distance while still offering filtered views from up close. Thus, a row of bamboo can function in the same way as a scrim is used in theater, only here instead of directional stage lights it is distance of the viewer that determines opacity.

Bamboo occupying the void between main building and perimeter walls at Nasher Sculpture Center, Dallas, 2003, as viewed from within and outside the main garden space.
[Photographed 15 years after construction]

A more practical application of bamboo can be seen at the **Nasher Sculpture Center** in Dallas (2003). Here, bamboo fills a void between the center's main building and the perimeter wall, in the process obfuscating fences that divide publicly accessible versus secured space. As a mass these plantings are effective at making visitors within the private site not cognizant of the nearby public sidewalk (and vice versa), an effect only amplified by the sound and visual interest these tufts of green mass provide with movement in the wind. As much as bamboo can be a signature element (as illustrated at IBM Japan Makuhari, for example), it also can be a surprisingly inconspicuous element of the landscape.

Bamboo linking spaces and managing to survive challenging growing conditions at the Center for Clinical Science Research, Stanford University, Palo Alto, 2002. [Photographed 16 years after construction]

Another use of bamboo can be found at the Center for Clinical Science Research, Stanford University, in Palo Alto (2002). This project's main atrium is exposed to the open air yet has the feeling of an indoor/outdoor room. Within these challenging growing conditions where planting areas are confined and sunlight reflects from immediately adjacent glazing, bamboo provide vertical shafts of green, linking the multi-storied structure and overhead pedestrian walkways while also offering some privacy to office spaces within. Despite bamboo's resiliency (indeed, it is difficult to imagine any other vegetation surviving these particular conditions), there are signs of damage brought about by light bouncing from the building facades and burning the foliage. Bamboo is hearty and versatile, but still has its limits.

AS SKIN

Any discussion of vegetation and landscape architecture would be incomplete without addressing the role of turf. Within Walker's work, turf can be seen as providing two main functions. First, the thinness of turf as a material makes it ideal for use when wanting to express topographic gestures. Second, in a way consistent with conventional uses but still worth highlighting, turf offers a tremendous degree of programmatic flexibility for supporting different forms of activity and intensity of usage. Where at least one of these concerns is present within a project, turf finds its way into Walker's landscapes.

Use of turf as a green skin that makes topographic manipulations visible was especially present in works from Walker's development-driven projects in the 1960s and 1970s. Examples include the previously discussed Foothill College in Northern California (1957–1960) and Sydney Walton Square (1968), in addition to early residential works in Southern California such as Baywood Apartments (1974) and **Mariner Square** (1970) that were completed for the Irvine Company. In each of these landscapes turf maintains a notable presence, but only where in proximity to pedestrians (that is, not as large vehicle-oriented expanses), and typically in some setting where that ground plane is guiding the movement of people and stormwater. We can recognize a scalar relationship between the earthworks and the surface material, where the (clipped) short height of the grasses allow the topography to have impact even when finely scaled. One can imagine how surfacing these areas with shrubs or even groundcover plantings would result in an entirely different effect where the underlying topography is less visible.

Physically, these turfed landforms have aged notably well and visiting them many years after their creation it is clear that they continue to create inviting spaces for users of these active landscapes. And yet, from a contemporary perspective they do look somewhat dated, and with good reason—the environmental costs associated with irrigating, mowing, and fertilizing these lawns make them less likely to be repeated today. These negative impacts can all be partially mitigated (such as using recycled water for the irrigation), but in the end they will continue to be resource-intensive, monoculture plantings. Generally speaking, turf allows for highly legible landforms, but this comes with environmental costs that are increasingly important to address.

Use of turf to accentuate landform, guide circulation, and maximize space within **Mariner Square**, Newport Beach, California, 1970.
[Photographed 49 years after construction]

Nasher Sculpture Center, Dallas, 2003. [Photographed 15 years after construction]

Where the use of turf not only continues to hold a presence but in fact spurs on continued innovation is in situations where the surface of a landscape will be actively used. A dramatic example is found at the **Nasher Sculpture Center**, whose sculpture garden is positioned upon a (gently tilted) ground plane of turf. As a design element, the greenness of this material helps reinforce a sense of oasis that makes it preferable to hardscape paving options for this sequence of outdoor rooms. At the same time, this downtown-located museum can have a steady stream of visitors in addition to hosting special events that together could easily overwhelm conventional turf plantings.

As a response to these challenges the designers worked with specialists to specify engineered turf and structural soils that can withstand large numbers of pedestrians. The turf itself in this case embodies multiple challenges of landscape design: supporting an urban level of uses while maintaining a verdant appearance; requiring significant technical knowledge and considered implementation, yet most likely being taken for granted by those who use the space. This landscape requires irrigation, equipment for which is housed within multi-use plinths that also hide lighting, security, sound and other paraphernalia while providing surfaces for seating and the display of sculpture. The grass plane at Nasher illustrates how plantings and the soil mixes that support them can be surprisingly high tech.

In the end, what can we learn from the use of vegetation in Walker's portfolio of built works? Most importantly we can recognize plantings as being key elements—the most crucial elements—made use of by designers in defining outdoor spaces. This spatial definition is itself a key operation for making landscapes more legible and potentially desirable. In Walker's projects we see examples of this definition being used to carve out a more human-scaled place (the willows at IBM Solana, for example) just as often as being used to maximize the use potential of an urban environment (such as at Nasher Sculpture Center).

Vegetation is also the landscape component with the greatest potential for improvement over time. Nearly all plantings within a built landscape will require some form of maintenance, but especially with trees, bamboos, and more substantial shrubs, time is on their side. The benefits of growth over time can be seen in showcase projects such as Saitama Plaza in Japan (2000), but also in projects like Boeing Longacres Industrial Park near Seattle (1994) where many paving and site furniture elements have largely disappeared even as key plantings are still thriving.

We can also recognize in Walker's landscapes clarity of thought in matching planting selections with specific landscape uses and activities. From this perspective, the planting design can be a direct response to more overarching conceptual goals. This might happen most typically

A verdant plane of turf at **Nasher Sculpture Center**, Dallas, 2003, lures visitors outdoors and is engineered to accommodate high levels of foot traffic.
[Photographed 15 years after construction]

in parallel with other site planning and design efforts, or indeed at times after the main elements of the site have already been configured. What is far less likely to happen is that the schemes will begin with the planting selections themselves. Returning to Nasher, the thinking is not that a collection of large magnolias would be beautiful and an opportunity ought to be found for their placement. Instead the focus is on recognizing that the main garden space will benefit by being inwardly focused and shielded as much as possible from the looming presence of nearby building towers, and with this goal in mind the magnolia trees are seen as an effective and preferable means of establishing a green wall.

Regarding the role of the landscape architect, in Walker's work we see the importance of not just planting knowledge but true expertise. Those outside the discipline may (understandably) expect planting knowledge to be squarely within the domain of landscape architecture, but in truth there continues to be tension for whether designers ought to somehow rise above such mundane considerations and effectively outsource this expertise to other specialists (i.e. botanists, arborists, or even just landscape contractors). Increased scope of possible expertise also contributes toward this distancing, exemplified by university curricula where landscape students might spend just as much time learning scripting computer programming languages as about botany.

For Walker's work at least, plant knowledge continues to be an essential and empowering resource for not only improving the work produced but also defining one of the landscape architect's core offerings.

In a more cursory review, Walker's use of vegetation can be seen as limited; since they generally do not celebrate change or ecological performance, the works are in fact static (so the thinking might go). But this would be a serious misreading of the portfolio, for two main reasons. First, we know that no landscape is truly static, as reminded through garden traditions such as bonsai where maintaining even the effect of stillness, of "the infinite now," requires tremendous effort that still is little more than an illusion. And second, in Walker's landscapes we find a distinct legacy of longevity and success in material selections, most especially with the plantings. Such success does not come by accident—it requires both knowledge and respect for the needs of these living materials, in their growing conditions, supported uses, and maintenance over time. This is the legacy of Walker's approach to "refining nature"—it is not about resisting these natural processes, but rather working with them and guiding these processes to produce a desired result. The value of those results ought to always be up for discussion, and the degree of sophistication in these planting schemes can always be heightened as has been the general trend in Walker's work (mostly for the better). But these judgments never have a chance to be made without understanding of the plantings themselves, because the consequential failures would simply lead to erasure.

Toyota Municipal Museum of Art, 1995. [Photographed 17 years after construction]

VII

WATER: THE DYNAMIC ACCENT

Water holds special roles, accenting, animating, and at times even defining a landscape. When thoughtfully conceived and skillfully implemented, water features can elevate green spaces to be more engaging environments.

"Water feature" is used as a term of art within the discipline of landscape architecture to encompass a wide range of elements, varying in scale and also kinetic properties. Misting jets, pop up fountains, water walls, reflecting pools, ponds, lakes. Larger water features are likely to have a direct relationship with topography especially if offering any performance value in relation to drainage, with (bio-)swales providing conveyance toward retention or detention ponds. For most urban sites, however, there more likely is an intentional separation between these stormwater systems and more highly controlled water features that in the end are offering aesthetic[1] (but note, not only *visual*) benefits.

Water features also present significant challenges of implementation and ongoing maintenance. For active fountains there is required equipment such as pumps and filters carrying financial and environmental costs. The physical space required for these elements can be significant, especially relative to the compressed hardscape spaces where these features are often located. There also is engineering of the system itself, sometimes developed in direct collaboration between landscape architect and engineer, but increasingly likely to fall within the more specialized domain of fountain designers.

All of these factors combine to make the landscape architect's role complicated: at once advocating for the benefits brought by these features, but then at times needing to caution eager clients who may not be cognizant of the significant commitment required; holding strong opinions about the form and effect of these features, but ultimately handing over final engineering to others.

Despite these varied and overlapping challenges, water features are conspicuously present in a high percentage (the majority, in fact) of Walker's landscapes. This prevalence brings an opportunity to better understand different roles such features can play in the built environment while also identifying approaches within Walker's own career that empowered him to see so many fountains through to construction. Viewed in the context of other 20th century landscape architects, Walker's use of water as a design element is distinct for its elemental quality and unique mixing of experimentation in expression with restraint in form.

Water feature as activator at **Nasher Sculpture Center**, Dallas, 2003. [Photographed 15 years after construction]

ACTIVATOR

Both visually and audibly, water can function as an *activator* of outdoor spaces. At the northwestern edge of **Nasher Sculpture Center** in Dallas (2003) a pair of rectilinear pools mark the edge of this site's extended garden room. As an attraction the water rewards visitors who have traversed the sculpture garden's lawn, with a configuration of framing pathways that invites continued movement around these pools rather than forced retreat. As voids these pools also establish a distancing between sculpture and viewer, while a simple row of gurgling fountains adds visual and audible interest to this otherwise very still space. These combined factors of circulation and spatial distinction revolving around an activated water feature make for a heightened viewing experience.

A related example of a fountain performing as activator can be found at the entry court of the **Toyota Municipal Museum of Art** (1995). At the center of this small, square parking lot, a circular fountain is defined by a low stone curb and functions as a roundabout for vehicles entering and then exiting the museum grounds. Jets pointed inward from the fountain's perimeter have a turbulent effect and echo the bubbling upper fountain that most visitors will later encounter at the extensive pool nearby.

Considering the quiet context and modest activity level of the museum, it is common to arrive at this entry with no cars parked or other individuals in immediate view. Under these conditions the tranquility of the space is punctuated by the movement of the fountain, letting the space still feel inviting even if desolate. Conversely, when programmed to host special events such as a weekend public market, the fountain is naturally embraced as a gathering space. The fountain together with material choices such as enlarged aggregate asphalt paving and fine edge detailing transforms this functional vehicular support area into an inviting plaza that contributes to the site's overall composition.

Water feature as activator of open space, both during special events and the more typical calm operation at **Toyota Municipal Museum of Art,** 1995. [Photographed 23 years after construction]

THRESHOLD ACCENT

A bullet train ride north of Toyota City and beyond metropolitan Tokyo, the water features at Saitama Plaza accent pedestrian transitions from the terrestrial level to the elevated plane of this "forest in the sky" public space. Washboard-like ramps of dark stone are offset from adjacent stairways, with the gravity-fed water trickling down the extended surface in a repeated sequence of rivulets. The feature is delicate in detail but dramatic in scale, focusing on a single water effect that is repeated for the whole of this tilted plane. Rectangular pools function as fountainheads at the summit of the two water ramps, and when filled provide reflective upward views to the surrounding forest of zelkova trees. In a detail iterated on in many of Walker's water features, the visible basin is lined with fist-sized stones that provide texture while also allowing the base of this pool to be presentable when emptied of water. The cut granite stone of the ramp is pleasant even when not in use, supported by the dark coloring and subtle modulation of each segment. All of this is framed by folded sheets of stainless steel that provide a crisp contrast to the surrounding natural materials.

The water features at **Saitama Plaza** are, ultimately, inessential, and in a different context under a different management regime it would be easy to imagine these features being vulnerable to abandonment. But in this finely crafted and clearly well-maintained site the fountains continue to function as accent elements that are valuable with or without active water flow.

Water feature used as accent to mark grade transition from street level sidewalks to upper level spaces at **Saitama Plaza**, 2000. [Photographed 15 years after construction]

Linear water feature as iconic axis at **University of Texas** at Dallas, 2011.
[Photographed 7 years after construction]

ICONIC AXIS

More than an accent or activator, a water feature can reach a scale where
it is a central, even iconic element in the landscape. This is the case at
the campus landscape enhancement for the **University of Texas** at Dallas (2011) where an extended sequence of pools forms a central axis that
is paired with a bold double allée of magnolia trees and links a range of
university facilities. This is the kind of element that in plan view can
appear quite basic but in *experience* has a dramatic effect. Part of this is
due to the feature's sheer scale, where it can take the visitor a good fifteen
minutes of walking to traverse from one end to the other (meanwhile,
maintenance staff make use of small all-terrain vehicles). Context also
matters, and in this arid climate the pools stand in appealing contrast to
their surroundings. The understated nature of the design plus its well-
executed construction make this a world-class landscape space within
an otherwise somewhat underwhelming campus. The shallow depth and
flat basin of the pools creates an inviting plane for students to splash in
on warm days, while short-reaching bubbling jets animate this surface.

PLAN-ORIENTED VOID

The shallow pools within the **Children's Pond and Park** at the Martin Luther King Promenade in San Diego (1997) present an illuminating counter example to the pools at UT Dallas. This water feature looks perfectly desirable when viewed in plan, with geometric forms anchoring one half of this public park along a major waterfront thoroughfare. But when experienced in person within what is a surprisingly isolated and challenging urban space, these pools feel highly out of place. The visual contrast between these vibrant elements and their more mundane surroundings could under other circumstances be playful, whimsical even, but when overlaid with evidence of homelessness and neglect the scheme seems inappropriate. Extensive signage declaring that the fountain is not to be entered contributes to the water feature functioning as a barrier within the landscape, while a stepping stone pathway crossing the main pool is too narrow to allow those walking in opposing directions to easily pass each other. Finally, an arced fragment of blue that viewed in the design drawings seems important to complete the pool's form, when viewed in person is more obviously separated by rail tracks and clearly unnecessary. In the end, this project presents some of the risks brought by the use of broadly scaled water features that are not actively programmed for play. This site also raises questions for how to judge, or at least interpret, the performance of a project that turns out to be largely unsuccessful but could be more highly valued in a different physical or cultural context.

The water feature at **Children's Pond and Park** at the Martin Luther King Promenade in San Diego, 1997, is out of step with its urban context, even if the project does still photograph well.
[Photographed 20 years after construction]

Water wall adjacent to cafe seating at **Nasher Sculpture Center** in Dallas, 2003, refined in effect but still bold enough to help define a social atmosphere. [Photographed 15 years after construction]

SOCIAL ATMOSPHERE

One of the most effective uses of water features is in creating an environment conducive to socializing, a place where small groups are naturally drawn to conversation. At the **Nasher Sculpture Center** in Dallas, a row of jets shoots water at a steady flow from a stone-clad wall, falling to form a solid splash line on the surface of a shallow basin. The fountain is *inviting,* if not dramatic, as the dozens of arced streams glimmer in reflected sunlight and have the kind of ordered appearance fitting to this gallery setting. Even more important than the visual appeal, the splash of so many streams adds up to an audible volume that permeates the adjacent cafe seating area. With a sound that is more akin to a stringed quartet rather than a single amplified guitar, the presence of the sound is enough to provide some privacy between the different tables of diners while not requiring individuals to converse at raised volumes. Positioning the seating area so that it in effect is inserted *into* the water wall means that the fountain splashes on two sides of the platform and creates a more immersive, surround-sound experience. Ultimately, this area highlights how the selection of a water feature type and its consequential effects ought to be directly related to adjacent program uses and activities.

Sometimes this linking of a water feature to social activity is a bit less direct, with the fountain acting more as another entity contributing to surrounding conversation. This is one way of considering the sequence of jets embedded within the ground level of a terraced gathering area at the the campus landscape enhancement for the **University of Texas** at Dallas. In built landscapes, there always is a balance to be found between needs and wants, and in this setting it is easy enough to imagine the planted terraces still being an inviting place for different configurations of students to gather even if the fountain was never added or indeed was cut through some process of "value engineering." But the space is improved by the fountain, breaking the silence of what might otherwise appear as a formal space. The fountain also offers a focal point, and with detailing that has the jets projecting through an extended drainage surface it begs to be traversed barefoot on a warm day.

Jets projected through a walkable platform and set within a gathering space at **University of Texas** at Dallas, 2011.
[Photographed 7 years after construction]

The effect of water as a *plane* is expressed at different scales within **South Coast Plaza Town Center** in Costa Mesa, 1991, and the main water feature at Toyota Municipal Museum of Art, 1995. In Costa Mesa, the fountain appears still from a distance and only reveals its minute movement upon closer inspection. Consisting of two thickened bands of water radiating from a central pool, thigh-high sheets of stainless steel are wrapped to hold the volumes of water, with precise edges allowing for consistent sheet flow down the face of the fountain even as its surface remains perfectly flat. The particular sizing contributes toward this feature's lasting impact. If it had lower height or was embedded into the ground, this water would be demoted and overwhelmed by the surrounding open space and adjacent building; if taller, it would be the sides of the basin that were emphasized. As it stands, the fountain's proportions give prominence to the water surface, views of which show some combination of reflections of overhead surroundings and refractions of the rough stone at the water's base.

Visiting this water feature decades after its completion, it is striking to see just how well it has endured and how elegant a site element it remains. Its weathering is helped by the generally accommodating coastal desert climate of Southern California in addition to the strict maintenance regime in place by the fountain's owner, but the ultimate legacy of this water feature is an example for how the elemental use of materials can allow each component to perform to its fullest potential.

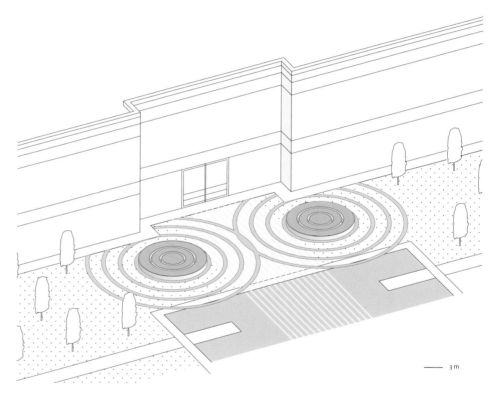

3 m

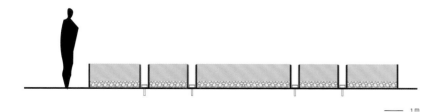

1 m

South Coast Plaza Town Center, Costa Mesa, 1991. Elements of stone, metal, and water each express their essential material qualities. [Photographed 28 years after construction]

Water feature as interrupted plane at **Toyota Municipal Museum of Art**, 1995.
[Photographed 17 years after construction]

The play on surface and mass takes a different configuration at **Toyota Municipal Museum of Art**, where an extensive and shallow pool maintains a still surface save for a perfect ring of fine bubbles making a precise interruption to the water's surface. The effect is suggestive without being metaphorical—is it volcanic? A platform about to arise? A deep sea creature? Clear answers are not necessary.

Awareness of the plane is amplified when approaching this space from either of the perimeter ramps connecting to the lower entry court. The fountain is completely out of view when starting at the ramp's low point, but as the visitor climbs there will be a moment when her sightline matches perfectly with the water's surface, no matter what her height. At this moment the whole of the fountain is a compressed line on the horizon with the aerated ring an outsized disruption to the surface.

CONNECTOR

An iteration on the idea of water feature as ornament is found in the **Forum 1 Courtyard** at the Novartis Campus in Basel (2004), where a linear reflecting pool not only provides visual interest at the heart of this courtyard but, even more importantly, acts as a link stitching the birch grove on the north side, a place for socializing, with the more formal turf area to the south. Warm-colored stone within the basin makes the most of this water feature's shallowness, while the use of precisely crafted stainless steel along the perimeter creates a fine infinity edge for maintaining an undisturbed water surface. The water feature clearly shares a lineage with the fountain in Costa Mesa, but here the same essential elements are reconfigured for a more compressed yet tranquil setting.

Linear water basin linking two courtyard spaces at **Forum 1 Courtyard**, Novartis Campus, Basel, 2004. [Photographed 14 years after construction]

Water feature as dynamic field at Tanner Fountain, 1984, within the Harvard University campus.

EPHEMERAL

Two additional water features from Walker's portfolio must be examined, as they represent a form of "min-max"[2] (to borrow Richard Haag's terminology) for how these elements can be made use of as ephemeral components in the built landscape. With Tanner Fountain (1984), water is expressed as a dynamic field rather than singular plane, an amorphous and changing element that is contrasted by the solidity of stone. This fountain—*this installation*—sits at a crossing of multiple pathways within the Harvard University campus. Rather than create an additional set of edges, this feature acts as an overlay that stitches together multiple elements of the site. A small clustering of trees and the backdrop of the nearby science center complex provide spatial definition that helps this low-lying feature not feel overly exposed. This setting and configuration, together with the scale of the feature itself, foreshadowed the fountain in Costa Mesa but are used here in a more engaging manner with greater spatial ramifications.

The dynamism of Tanner Fountain is expressed at multiple time scales, with change from one moment to the next but also across seasons. In summer, gentle jets shoot water waist-high and offer playful relief from the heat. In winter, the original design released a cloud of steam, being a kind of "anti-fountain." Both of these modes offer the design lesson how many fine jets will create a larger whole that can be read as an elevated volume. And perhaps even more lasting, there is a lesson for how a field of elements can overlay a relatively complex set of circulation conditions as an alternative to simply subdividing a given area. Even as these very lessons have been overlooked in more recently constructed areas immediately adjacent, the legacy of Tanner Fountain persists.

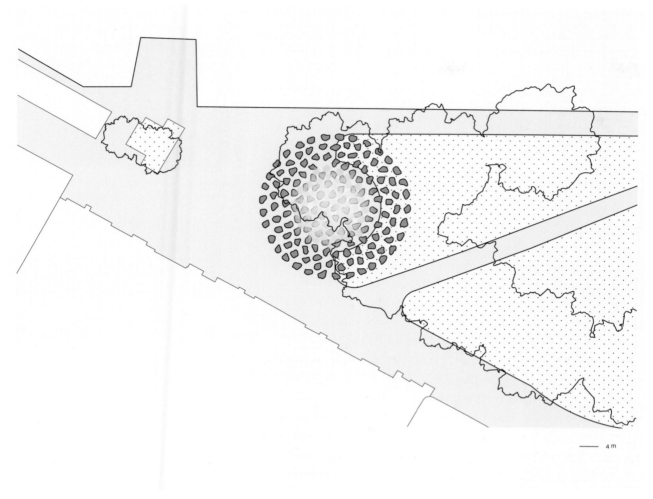

— 4 m

Tanner Fountain, Harvard University, 1984.

The water feature at **Jamison Square** (2006) both divides and unites its city block-sized site. Beginning with a flat plane, the wall runs north to south with a basin carved out of the western side to reveal an exposed face of the wall while creating a depression for an ephemeral pool. The result is two landscape "rooms," with the eastern side offering shaded spaces for gathering and the occasional display of sculptures while the western half is sought out as a place for socializing and play.

One of the most effective qualities of the water feature at Jamison Square is the way in which this element not only offers an attraction that is pursued out by park visitors, but this same feature also has an atmospheric effect in shaping the environment and "owning" this space. The feature exemplifies an aspiration of design where a single element provides multiple functions, sometimes in surprising and not entirely obvious ways. Sound is a particularly important component here as can be the case with water features in general in urban parks.

To most visitors of Jamison Square, it is the interactivity of the fountain that leaves the strongest impression. Not just allowing, but supporting and safely encouraging active use of a fountain requires significant design consideration. For example, here water emerging from the wall is substantial in volume but the pressure is never strong enough to risk sweeping away a child. When fully activated, the hemispheric basin fills with a pool of water deep enough to be stomped through by a child, and yet shallow enough to minimize risk of drowning should there be a fall.

At different scales and to varying effect, Tanner Fountain and Jamison Square offer lessons in how the ephemeral dynamics of water can be embraced to create spaces that engage activity.

Jamison Square, Portland, 2006.
[Photographed 11 years after construction]

For all the challenges associated with water features, Walker's portfolio presents a potent argument for why these elements are worth striving to have then become integrated into our built landscapes. The key offering is a high level of versatility that leads to the features capably fulfilling a wide variety of roles, as illustrated above. The fountains themselves can take on many forms, but when playing off other design elements this distinction can be further amplified. The thoughtful integration of a water feature can have a transformative effect on a landscape.

The surveyed projects also suggest that these features need not be grand to still have impact, especially when complementing or accenting other landscape features. None of the designs could be described as entirely dependent on these features, and even for Jamison Square or Tanner Fountain there still would be something of value and interest for visitors should the water be turned off. As is the case with so many other components, most important is that the features are well-considered and intentionally employed.

With water features requiring such significant commitment of upfront investment and ongoing maintenance, their use indicates a client taking the project seriously. In this way the preponderance of water features in Walker's portfolio gives us, in at least a cursory way, a sense for how consistently he was designing for the upper realm of built landscapes. Most landscape architects do not have the (earned) opportunity to invest so much in the iterative advancement of this landscape element. Evidence of this empowered role is one of the most lasting legacies of Walker's career.

1 As a reminder, aesthetic is defined as "concerned with beauty or the appreciation of beauty."
New Oxford American Dictionary, 2010

2 Thaïsa Way, *The Landscape Architecture of Richard Haag: From Modern Space to Urban Ecological Design*
(Seattle: University of Washington Press, 2015), 77.

VIII

CONCLUSION

This study has been predicated on an assertion that built landscapes hold lessons of design best identified through in-person investigations combined with focused desktop study, and that these interpreted legacies of existing landscapes are valuable for informing design responses to current challenges—that history has something to tell us about today, tomorrow.

The detailed strategies outlined above have intentionally been stripped bare, with the intent of being as accessible—as *useful*—as possible for a current generation of designers seeking to incorporate these lessons into their own efforts. While this structuring has allowed us to take on an expansive body of work in a way that produces specific insights, there also are more comprehensive reflections worthy of consideration.

What are the broader lessons we ought to take with us after completing the journey of assessing such a singular portfolio?

PHYSICALITY AND CONTEXT

Experiencing Walker's landscapes in person made each of these projects specific rather than generic or abstract. The works became real in a way so that even now as these words are written my mind shifts to think of how these projects might be faring at this very moment. Presenting the projects as has been done has hopefully made these works more real to the reader as well.

Experiencing the projects in person revealed their physical limits, bounding the landscapes but also providing illuminating and often flattering contrast. Surrounding contexts were necessarily interacted with before and after exploring the study sites, prefacing and then in their own way offering epilogue to the experience of Walker's landscapes. The immediate juxtaposition highlighted the value of nuanced elements within Walker's landscapes that might seem insignificant through imagery but in experience are precisely what elevates these projects beyond the everyday.

For example: after a jet lag-induced sunrise visit to IBM Solana, I made the relatively short drive to downtown Fort Worth in pursuit of Burnett Park. With the highway acting as a transect, there was an expanse of rural and exurban landscapes traversed before reaching the city center. Transitioning to foot, crossing an oversized and heavily crowned street, passing beneath canopy cover and entering into the park—the space felt impossibly fine-crafted, a jewel of a landscape within a more mundane context. How far, I wondered, would one have to travel before finding the next example of such a landscape? Perhaps it would be Walker's own Nasher Sculpture Center in neighboring Dallas.

These experienced contexts can also be humbling. For example, between two visits to Barangaroo Reserve, a brief hike along the nearby natural shoreline made clear that even a work as ambitious and well-implemented as this one could never—ought never *try*—to directly compete with the real thing. In this way the scale and dramatic beauty of the coastline between Bondi and Coogee highlighted how essential it is for a built landscape like Barangaroo to support activities and experiences distinctly well-suited to its urban community. Meanwhile, passing by the adjacent gleaming office towers as workers headed out for happy hour called attention to the significant public value of the park vis-à-vis vanity architecture housing the one percent.

ADJACENCIES AND TYPOLOGIES

Seeking out these works with the benefit of hindsight and knowledge for how each fit within Walker's own career timeline provided a kind of temporal context to the projects. For example, walking the inviting grounds of Foothill College, feeling the cohesiveness and integrity of the scheme still evident today while also knowing this was just the very start, that there were scores of projects to follow—it emphasized how Walker's own landscape skills and sensibilities were there from the beginning.

Another example: touring the grounds of IBM Santa Teresa in San Jose, California, right as this manuscript was being finalized, I was impressed by the care with which this site has been maintained and the way it still so boldly expresses its original design, sure. But even more so, while exploring the campus I kept thinking about how this project was completed at a moment of consequential transition for Walker in the 1970s, right as he was leaving California to begin teaching at Harvard, and with that move was intentionally stepping away from overtly commercial and developer-driven projects to pursue institutional clients including IBM itself.

Finally there are the projects that started off this study in the first place, the collection of works in Japan. To me these sites collectively represent a high-water mark for craft in landscape making, and although all have been in place for many years, it was only subsequent to initial visits that I recognized most of these were completed when Walker was already in his sixties. At this stage of his career Walker had acquired multiple decades of professional experience in addition to many years of intellectual advancement through teaching and writing, establishing a foundation especially well-suited to these Japan-based groups of clients, builders, and collaborators. Return visits to Toyota Municipal Museum of Art in particular called attention to how any built work emerges from a specific space-time; for this finely conceived and implemented landscape everything came together in a way that could never directly be repeated.

AMBITION AND LIMITS

Walker's influence reaches far beyond his landscape designs themselves. Scores of associates have passed through his firm, many of whom who have gone on to establish thriving practices and even teaching legacies of their own. Through the years he has had hundreds of students, and with his distinct efforts in publishing Walker has established a set of resources that continue to be made use of in universities and design studios.

Still, it is the built landscapes that are the ultimate ends, ends offered up to the world as physical spaces that create their own opportunities. And while the soft influence of Walker's career is inherently unquantifiable, the collection of built works very much can be tallied—I have done it. Such a list spurs ambition for its scope and consistency of output across so many years. Indeed, most designers would be lucky to produce a quarter—ok, maybe a tenth—of the high-quality projects within this portfolio. And yet when those projects are *spatialized* and mapped, zooming out shows the works as infinitesimally small patches scattered about the (digital) globe.

Visualizing the limits of one career is humbling, but also clarifying. Applying our skill and energy toward contributing a few more of those small patches, serving our communities and their environment as best as we are able to—perhaps that is what it means to be a landscape architect.

IX

EPILOGUE:
IN CONVERSATION—
INTERVIEW
WITH
PETER WALKER

This interview between the author and Peter Walker explores a range of intertwined topics, personalities, and reflections on the designer's career. The discussion provides an additional frame for interpreting the legacy of this portfolio while also making use of Walker's unique perspective to comment on broader themes concerning the discipline of landscape architecture. The transcript has been edited for improved clarity and concision. The interview took place on 22 February 2019 in the office of PWP Landscape Architecture, Berkeley, California.

SCOTT MELBOURNE I have prepared some questions for this interview but am happy to jump right in. You were talking about how for designers like Rich Haag politics were at the core of his interests.

PETER WALKER Yes, but they were hippy politics. They were anti-establishment politics. Meanwhile [Hideo] Sasaki saw this postwar growth and was trying to figure out how a landscape architect could command that growth. Universities, highways, shopping centers, housing—all those things were exploding. And [Donald] Sakuma and others, all of us at Harvard in the late 1950s, each in our own way picked up that enthusiasm from Sasaki. Rich didn't really share that. He saw the growth as potentially dangerous, and in some ways they were both right. I mean you had to grab the tiger and try to ride it. But at the same time it was sort of unridable. And now as I drive around and see, not just in the United States but also certainly in Asia—it was more than landscape architecture could handle. You have to remember there was this other field of environmentalism, which was Grant's [Grant Jones] basic thing. Because [Ian] McHarg was huge at that time, and he was a competitor with Sasaki. McHarg said you've got to start protecting the world from this growth, humans are no damned good. And Rich was really kind of like that. He wasn't McHargian, but he picked up the same kind of skeptical thing.

I remember McHarg used to tell this story: After a neutron war, there were these two amoebae down in this pond. One amoeba looks at the other one and says ok, this time no brains.

SM [Laughter] Right.

PW That was McHarg. I mean, he overstated tremendously, and while that point of view didn't exactly win, it clearly did institutionalize itself. Capitalism was just too much. And the art side—my side—was like it always is. You're dealing in icons, you're not dealing in solutions.

SM I have traveled the world visiting your projects just trying to learn from them. You might remember the first time you were kind enough to meet with me. I had just visited your projects in Japan and thought I might write something up about those, and now it has grown into this larger book. One thing I see as so especially valuable about your legacy is how you fulfilled really the maximum potential of a form of practice.

PW Well, to be fair, we did it—Sasaki invented the corporate practice.

SM Which you have written about.

PW Sasaki wanted to bring all the environmental arts together. Sasaki Associates was his idea. And we did that for a while in California, we got up to a couple of hundred people. But we weren't producing icons. I had given up on trying to control the world, which Sasaki never really gave up on. I don't know what he thought in his older life, I didn't talk to him about it, but I just decided that was impossible. And I looked around and the two things that influenced me the most, partly because I got a chance to visit them, were the Japanese gardens of all sorts—imperial, Buddhists, whatever—and also French gardens. French and a little Italian. Because they had produced icons. They were my goal, and my model. Mostly [André] Le Nôtre.

One of the things about modern landscape architecture is that it's *big*. We're not doing little Ryoanjis. We're doing de Sceaux, Versailles and so forth—that's the scale of what we do. A campus, a shopping area, housing complexes, civic complexes, memorials. That was my point of view—and I've never suggested that was either the only or the best. It was just mine. And I just stayed with it.

I first spent a lot of time in France and Italy, then in the 1960s picked up this art thing because artists were working at an industrial scale. They had shown how you could get outside, how you could do big things. And for me the minimalists were always better than [Isamu] Noguchi, because Noguchi was at root a sculptor. I mean I really appreciate that, don't get me wrong, but I didn't think you could sculpt the world.

SM I grew up a bike ride away from California Scenario and was kicked out of it as a kid for playing in the water. [Laughter] But your description of Noguchi as at the core being a sculptor—we can see how that sets the tone even in some of his most amazing works.

PW Well, he did other things. He was a sculptor, but to the degree that he was a landscape architect and talked about bones, the spatial bones, he meant sculpture. It was not about planting, it was not about irrigation, it was not about growth. It was fixed.

I saw a wonderful show of his in Washington a year ago, which comprised sculptures I hadn't seen except in pictures. They were all in this show and you could see the development of his thinking. And you could also see the limitation, if you looked at it from a landscape perspective. He wasn't interested in landscape.

Le Nôtre's inspiration was agriculture. I suppose that's a California thing as well. It's a Midwestern thing too, but in California you can see it. You go up to the vineyards and you can see that design. You go out into the orchards, and you go out to the annual crops, and that's what Le Nôtre was looking at.

SM Do you see agriculture as also empowering, because of the scale?

PW Exactly. The scale is *empowering*. And the geometry, play it against the scale. It doesn't preclude you thinking about water. It doesn't preclude you thinking about soil science. Which, interestingly, Rich knew a lot about—more than most of us. But he never used that as the basis, and of course his teacher, and Sasaki's teacher, and my primary teacher was Stan White. He was about the way the earth was.

SM It is fascinating to think about that history—Stan White, Hideo Sasaki, Rich Haag, and yourself—and the distinct directions you all took.

PW In my own mind I make a real distinction between Sasaki and SWA. SWA took on housing, which Sasaki was never very interested in.

SM That's interesting, right.

PW And it's interesting because at root, when McHarg was at Harvard he was interested in sociology not earth science. He was interested in the new towns in England. And I got interested right out of Harvard because I had done some housing in school. I got interested in the new towns. It was the Finnish, Swedish, English, Scottish, some German, new towns. And that really was the twist that SWA had but Sasaki didn't have. Because there were all these developers and no landscape architect—or architect—was working for them. Almost none. So it was possible to form a whole new genre.

SM Which you did.

PW Which we did.

SM Just a few days ago I went back to Mariner Square, believe it or not. I'll tell you as someone living in Hong Kong, to walk through the gate and enter into a compressed courtyard—and seeing the early days of your relationship between topography and circulation, how that also maximizes space, and to see then of course the shared pool—it still today feels so incredibly livable. Not decadent—inviting.

PW The scale certainly came from the new towns, but there also was a growth of a kind of social idealism. You find early examples in Long Island by people like Albert Peets and Stein and Wright, who took that idealism and built projects just before the Second World War. They were forming firms that influenced landscape architecture across the United States, particularly in the Midwest and in the Northeast. And then the Second World War happened, and it died. Just like parks died. Just like universities had a kind of lull. And then the work picked up and ran so fast we couldn't catch up.

Most people, most designers are more business-oriented than they are artistic oriented—or intellectual for that matter.

SM Or maybe it's the business-oriented designers who survive.

PW Yeah, or thrive. I think of the architects from just after the war and some also prewar like Skidmore, Mies van der Rohe—there were a number of architects who were very serious about the social commitment, and how you made places that people could do things in. And there's not much of that now. They've all become commercial. Skidmore, Owings & Merrill is now a commercial firm, not really dealing in icons. The buildings are sculptures. And in their way they have the same limitations I always thought Noguchi had. Because he had no social sense—he could hardly carry on a dinner conversation. He was so angry.

I went to Italy to see this little town called Pietrasanta that made stone sculptures. Noguchi had two manufacturing places—Takamatsu in Japan and Pietrasanta. That's where—these were industrial stoneworks—you know everybody thinks he carved them out—nonsense—it was just like the Renaissance is all baloney.

SM [Laughter] So this would be the early 1980s when you were going there?

PW Yeah, 80s and 90s. There were factories that made stone sculpture. A close friend was working over there for Noguchi. And we bought a little place in Pietrasanta and I learned a tremendous amount there. You go through stages but you change—not necessarily for the good—but you change because of experiences. And by watching experiences, by seeing what works and what doesn't work. I mean by being analytic, and thoughtful, you can then grow.

In teaching I was trying to get people to think about spaces, to think about landscape in spatial terms and in artistic terms. But not just sculptural. We finally developed a class that started off taking ten contemporary sculptors—not just minimalists, but postwar from Calder on. Essentially doing papers on these people and on their work, trying to understand what they were doing. And the next step was to do a garden to be inhabited by one of these people, to understand the

relationship between what they were doing and what the space needed to do it. And then the next step was to do a memorial where you brought in the spiritual—social spiritual—and then using the tools that you had learned in the other experiments. And the results were remarkable.

SM Believe it or not a form of that exercise is still conducted at my university—I now know the origins.

PW Well, you have a set of tools. When I went back to teach at Harvard, which was in the early 1970s, I had pretty much given up landscape.

SM Yes, I want to talk about this! This is so amazing!

PW I had kicked myself upstairs at SWA and ultimately resigned. It'd got big—I was dealing with bankers, and insurance—you know it's like any corporation. The operation of any corporation is more important than the product. And I didn't want to do that. So I was going to become a sculptor. And I got a loft in New York on Crosby Street and I would spend my weekends there.

I was very enamored with Donald Judd, and very enamored with a number of those minimalists, and they were using industrial process, much the way, that Noguchi was. After seeing this incredible work I realized it was going to take me forever to learn—in craft terms—to be able to manipulate that stuff, those processes. And I got very discouraged. I was producing things but wasn't happy with them. And one of the things that occurred to me was that I'd spent twenty-five years learning how to make landscapes. So I went back and started making landscapes that were pieces of art. And that was where that came from. And the classes at Harvard, the thinking processes, that was the genesis of this office.

SM That phase in your life and career, I'm not sure how well-known or appreciated it is. But when I put the pieces together—we can learn so much from the efforts, the ambitions of others. A career is a kind of case study in itself.

PW That's right. If you play it that way.—

One of the things I learned when you deal in manufacturing—doesn't matter whether you're making a refrigerator, a car, whatever the hell it is—you decide what it is, you set up a Henry Ford thing, and you make a bunch of them. And you sell them everywhere. And if they're useful—not necessarily beautiful, but if they're useful and even better if they're beautiful—you essentially are pumping out those ideas into the world. Literally worldwide. Artists on the other hand are dealing at the root not with construction but with the ideas—it's the spirit of the thing. And if you're a painter and you live in Japan, or if you're a painter and you live in Italy, your paintings change. But landscape is even more localized. Climate system, a set of plants, the type of earth, culture—all of those things. So the landscape by nature can't just be exported like a refrigerator or a car, or a piece of furniture. It has to be engaged. So, in some ways, McHarg was right in that each piece of ground, each region is different. But what he was wrong about, I think, is that he gave no weight to culture. Surprising because he was a sociologist when he began, but he decided that people weren't any good.

SM That may have colored his philosophy—

PW And he purged it. He taught his students to purge it. And of course to do design in a world where you've purged the reason for doing it won't amount to anything—hasn't amounted to anything. At least at that scale—maybe at a regional scale it could amount to something. If you think about Los Angeles now having reduced smog, or having got a rapid transit system that's basin-wide, those are incredible things. But they're not design things; I mean they're not much to look at!

SM In the intro to the recent Reed Hilderbrand monograph you wrote that, "To successfully build a fine work of landscape architecture is one of the most difficult tasks in the world of design." That's

powerful to be said so directly, as I believe it is somewhat lost or at least not fully acknowledged in current discourse. But then coming from *you* of course it's even more powerful since you're not someone who's accomplished so much only to go on and say "Oh really it was quite easy."

PW Well, I think part of that difficulty is because you have living material.

SM Yes.

PW People don't realize how predictive landscape has to be. In that it's quite different from architecture. People wonder why something in architecture looks really good, it's because it's been rebuilt ten times—very much like Italy. They work at it until they get it right.

Landscape on the other hand, as it grows, it gets better, and then it declines. It's continuously declining at the same time as it's continuously growing. It's not just about predicting, it's trying to work against this inevitable erosion. I suppose in a way that makes it zen, where you have to accept.

When lecturing I often tell a story about falling in love with Le Nôtre's work, falling in love with his rows of trees. But the first time I tried it, it just looked like hell. They didn't grow straight, their heads didn't come in even, and I realized there was something more to it than just planting the things.

SM Let alone just drawing the things!

PW Yeah. And I started paying more attention to agriculture, to how you essentially *manage* something into existence. And that led to *who* manages, who keeps something and keeps it going. So I tried to find institutions because they had a longer term interest in whatever was being built.

I was talking to this guy at Skidmore, who was writing an article about Weyerhaeuser because the client moved out of the building. And I went up there when they were just moving out. And it was wonderful. The damned thing was forty years old, it was mature, I mean it just was fantastic. How could they move out of it, it was so much the image of their company? It was like a palace that came out of an agricultural basis, which is what Versailles is, and Vaux-le-Vicomte, and Sceaux. Using agricultural means to get these effects.

Now agriculture in the 17th century was a management thing, not sculpture. It was managing plants to make some kind of architectural thing. I remember when I first saw Sceaux, it was the only landscape I'd ever seen—I hadn't been to Japan yet—it was the only landscape I had ever seen that was as good as Chartres or as good as Notre Dame, or Salisbury. But Notre Dame and Salisbury are more fixed, and they also have an institution in them that keeps them going.

So, a great client of mine became IBM, also Weyerhaeuser, and Upjohn. And Martha [Schwartz], at the same time she was going after art people—none of her work still exists. Almost nothing exists, because the art people are fickle. And the other thing is they don't see the landscape as a work of art, which she does. But it's very hard because almost all of her better work is essentially gone. That's part of this erosion—part of the erosion is cultural. Because if you don't value it—like a car if you don't drive it, wash it, oil it, you know. And so we were looking for people who could do that, who would likely do that. That's why I think Weyerhaeuser was such a shock—if you're doing some housing project and it goes to hell you kind of expect it. People used to call me up, they'd say you know we're living in a cluster housing thing that you designed and it doesn't look very good.

SM Thanks for the call!

PW No, no, they said, could you come down and look at it. And I would because I was interested. So I'd go down there and all the hedges were leggy, the trees had branches falling off and so forth. Of course Le Nôtre's things have been built over, and over, and over again. They don't just exist. There's no three hundred year old Le Nôtre garden because trees don't live to be three hundred years—maybe a hundred, but not three hundred. They're replanted.

So I told the residents: Look, you really have to start over here. And that was the end of the consultation. It was too much for them.

SM This for me was an "aha" moment, recognizing how your focus on institutional clients combined with interests in agricultural practices and maintenance upon returning to Harvard, post-SWA.

PW When you move from architecture or sculpture to landscape, what are the differences? If you just go about it as a sculptor or go about it as an architect—and there are some landscape architects that work pretty much like architects. They lay it out formally, not use plants very much, or just use them as accents but the structure is not built with plants. You know, you're tempted, because it's fixed, it will stay put. But then it's not landscape.

SM After examining so many projects I feel I've never been more aware of the mortality of landscapes. And like you said earlier, in a way it is somewhat zen.

PW So. Let's say you get married and you have some kids. They're alive—going to die eventually—but going through all these stages. You're a parent and you love them and you care about them. So then what do you do with your kids to make sure that they at some level succeed? You think about education, you think about religion, you think about social values—you think of all the things you want to give them as tools for their future and their success. And the gardens can be thought about exactly the same way.

I think I told you at one time that Japanese people would come up to me and ask, "How did you do these Japanese gardens?"

SM [Laughter] Right, sure.

PW I remember a story about Sasaki. He was the only Japanese on the Harvard faculty—so when Japanese people would come to visit Harvard they'd call Sasaki up. Not landscape architects, just anyone. It was insulting, because Hid wasn't Japanese, actually he was American.

SM Yeah, he was born in California.

PW Yeah, on a farm—which he hated.

SM Ah.

PW So anyway, he would do these things and he would bridle. And then someone called us up—I was there at the time—and said, "You know we have this project in Japan and we want you to do a Japanese garden." I remember Hid, I heard him say, "I don't know how to do a Japanese garden." On the other hand, when we went there and started doing stuff, going around and seeing the gardens you're affected by it. So there is a cultural tie. But it isn't that you're a Buddhist or a Shintoist, you don't know that level.

SM You absorb the aesthetic.

PW Yeah, and the military aspect of their culture—all the shoguns and stuff, still there.

SM So much of your life and history is tied to California—how do you think about that at this stage?

PW Well, I think—Boston is too hot and too cold. [Laughter] What I've always loved about California, if you want snow you drive up there and get it—I don't like it delivered to the house.

But, I think a lot of what we are, what I am, is from the West. But we don't do Halprin-esque things, we don't do [Thomas] Church. I mean the root of our practice is so different from theirs. Different in time. We're probably closer to Church in that Church had this tremendous interest in Spanish history, because he thought he could connect it to California. Though I think his later works, particularly those that were influenced by [Lawrence] Halprin and art because Halprin was trained—he knew who [Joan] Miró was, he knew who Mondrian was, and so forth, Halprin introduced Church to art. And the later gardens, those sort of Miró-esque gardens are very similar to Noguchi because it was the same group of artists. And the central question of *Invisible Gardens* was, what were the influences? What made the people different?

I think the cultural differences probably have deeper roots than even the experiential ones. So I don't feel close to Halprin or Church. I mean, it was great to know them, they were intriguing people. But I don't feel very Californian. I'm more, not so much Boston—I don't know, teaching there for twenty years didn't really stick with me. But New York and particularly the art world in New York and by extension the art world all over the world, that has been more influential.

I remember going to see a show in New York of Richard Neutra's architecture, and somebody said, "If Neutra had lived on the east coast he'd be more famous than Wright." And I think he was a better architect than [Frank Lloyd] Wright, much more sophisticated. He's European for one thing. And was a modernist not in the prairie way, he was a modernist in a sophisticated European way, like [Charles Rennie] Mackintosh—the early modernists. So I think California gave him opportunity because people didn't know what he was doing, and he had a whole lot of followers. But then he lost out because of the market, and in particular the intellectual market.

I never suffered from California because we had an outlet for our ideas, both in terms of building things but also in terms of representation. Halprin had to write all his own books, I mean nobody ever wrote a book about Halprin. Church was a pretty good writer, better than Larry [Halprin]. But Church was discovered by *House Beautiful* and *Architectural Forum*, and thus became known.

That famous story where some Swede called him up and said, "Mr. Church I'd like to have you do a garden by my house." And Church said, "Where are you?" Well, he said, "I'm outside of Stockholm." He said, "Well, I've never been there and I don't want to come over." And the guy kept at him and said, "no, no, come on over." So he goes over and this guy was building a house in the middle of a birch forest. And Church looked at it and said—

SM Done!

PW There it is! [Laughter] You know he used to go up in the valley and he would sit in bad furniture on a bad terrace and sketch, do it for a hundred bucks. Give the client the garden design and say you build it.

Church wasn't philosophic. Halprin was philosophic, he was trying to breed meaning into things. Urban this, and social that. And I think the kibbutz—the agricultural part of kibbutz—was a big part of that. Because he had an agricultural degree, his first degree was science. And then he picked up the design at Harvard.

SM I am thinking about Halprin's legacy—a place like Freeway Park in Seattle.

PW And Rich worked for Halprin. I think Larry was unfortunate on several lights. He was very well educated. He'd had this experience that was quite worldly of living in a kibbutz, which is also very intellectual. But he was like an artist, he was less interested in other things than his things. Huge ego. You could run into him and he would be immediately angry and defensive. His major wins were on the west coast, both in terms of the early gardens but also these public spaces, which I think are his high point. And then, he got the Roosevelt Memorial in Washington. The whole history of the memorial is really strange. It went through three competitions, and Halprin was finally chosen simply because the jury was exasperated. But then they didn't build it, and twenty years went by and it was built pretty much simultaneously [1982] with Maya Lin's Vietnam Veterans Memorial—she was only twenty-one at the time. And everybody made a comparison between these two. I remember writing an op-ed, saying, that this was tremendously unfair and also ignorant of his work. Halprin had invented these ideas twenty years before, and he's been compared to something that a young person did—a great thing—but without this veneration of the time and so forth. He was also trying to incorporate the hippy lifestyle. He was like Rich, only he was an artistic hippy.

SM Rather than a spiritual hippy.

PW Yeah. I wonder, if we hadn't written that book [*Invisible Gardens*] about these people I would know less about them. I gained tremendously from learning even a little about them, and then trying to relate them to the art that they were aware of, trying to relate them to where they were, the sociology, etc. It was kind of like looking at a catalog. And one of the things about looking at a catalog is you make choices. You say I like this, I don't like that. It's like going to Italy or going to France.

Something I've thought about a lot is how do we get people to see? Not like a camera, but the perception of the essential things. I used to say to my kids and students, "If you don't find something you can steal and make use of in this museum, it wasn't a successful experience."

SM [Laughter] Right.

PW Picasso had a record of many, many previous artworks. He would go through it, think about them, and then he would appropriate and use those things. But no one knew he was doing it because the things he was making didn't look like those previous works—and I always thought that was smart.

It's the old thing about nothing new under the sun. But the real phrase should be there's nothing absolutely new under the sun.

SM *Invisible Gardens* to me is such a treasure—it marks a moment of an alignment between your own interests and those 20th century figures, their legacies still being fresh. How did that come about?

PW Well, I was teaching a seminar, and I gave each of the twelve or thirteen students an assignment to study a different contemporary or nearly contemporary landscape architect.

SM Good structuring, sounds productive!

PW So the students looked at this guy, came back with two pages and then made a little presentation showing pictures. Well—it was a disaster.

SM Why—because the results were off base?

PW Yeah, or just completely wrong. It was nonsense basically. And I knew enough about these people to realize this was just baloney. So at the end of the semester I said, "Where did you get this stuff?" "Oh, there were magazine articles and essays." And who made these articles and wrote these essays? Well, the designers wrote them themselves!

SM [Laughter] That's one way to do it!

PW So after a few semesters of doing that I brought Melanie Simo in because she was an art historian. She was helping the students do research. But we were not making much headway. One semester to the other it was still the same nonsense. And then she suggested to make a lecture course out of it. So Melanie and I would present instead of the students. Melanie brought an understanding of the art influence on each of these designers. What art did they know about, what were the ideas they were stealing and using? So we gave this lecture, and the lecture eventually went up to forty people because the architects were coming as well. So I went to Eduard Seklar, who was the senior historian at the GSD after [Sigfried] Giedion had died. And I said, "Eduard, I'm so sick of this lecture, I don't want to do it." And he said, "Write a book. You never have to give the lecture again."

SM When I was first digging into that book, I was thinking that it was something you had produced relatively early in your career. But then I realized you already were quite accomplished at the time.

PW It happened sort of in the middle. And I couldn't have done it without Melanie, she was instrumental in terms of asking questions, and when we didn't know the answers we had to keep going.

Speaking of publications, what is your idea of this book you are writing?

SM It is my attempt to make the knowledge and wisdom embedded in your built landscapes as visible as possible.

PW I think that's a very good way. Most landscape architects are not scholars. A few are historians who would do monographs that are scholarly. But most designers don't have the background, or the training, or the skill. The books produced often deliver a kind of objective description rather than looking at the ideas behind. We don't have very many people who are willing to give the time and the energy to think deeply enough.

I think two things in US academia are really unproductive. One of them is that practitioners have been virtually eliminated from teaching. If you're gonna teach, you're gonna teach. And the other thing is that the local groups of landscape architects want the teachers to teach rather than be in private practice and compete with them. Well, imagine if that were true in the past—there'd be no Rich Haag, there'd be no Sasaki, if it was that rigid. So I think those are real losses to the profession.

My wife [Jane Gillette], who is an English literature PhD, said if you're going to enrichen landscape intellectually you've got to find people who have intellectual discipline. And that's what we're very short of in the profession.

The other thing I would say to you is, I wouldn't worry too much about failure. While not all books are outstanding, none of them are really worthless because it isn't black or white and you just do the best you can. I also think you have to pontificate because there's such a big gap between knowledge and experience.

SM I feel my subject was a good choice in that you have just so much work out there. And it being landscape, it's accessible and ready to be experienced.

PW If you were in any other field than landscape architecture—if you were, say, an art historian—it would be required for your advancement that you would write, that you would bring your findings to at least a peer understanding, but hopefully a broader understanding. And design doesn't do that. A small exception are a few landscape architectural historians—but they write fairly narrowly.

SM Yeah, narrow and deep.

PW When I was chairman [at the GSD], one of my jobs was to monitor the classes. I would sit in on a whole semester and see most of the lectures of a historian, and we had great historians. What I think is so interesting about history is they're talking about the same thing but it's not the same story. I mean it's interpreted. And observations are really important. So that's why I say I wouldn't worry, just get it done.

SM Thank you for that. In a way I've found—talking about the limits of landscape—I've found the limits of a given book. I have a lot more to say.

PW Write another book. One of the reasons we got into publication was because no one was doing anything. That no matter how bad what we did was it was—

SM Infinitely better than nothing!

PW It was fascinating for me because a lot of times it draws you into thinking about something you hadn't thought about.

SM That's the beauty of writing.

PW Well, and it's also the beauty of looking. Of going around and seeing things.

SM I was asking earlier about your thoughts on your relationship to California. This place—especially in that last half of the 20th century—was probably the best place on the planet to be an active practitioner of landscape architecture.

PW And San Francisco was one of the best places.

SM Oh yeah, I can see that. While at the GSD I was in I think the last studio that George [Hargreaves] taught there. Our site was here in San Francisco.

PW George was very adventuresome, one of the best students I ever had. One of the hardest students, but—

SM [Laughter] I can just imagine.

PW He was one of those that caught on to this thing of utilizing art as a basis for design. I haven't thought about George in a long time. [Laughter] We had a jury on a residential project, a garden basically. And I invited Frank Gehry and Richard Serra to the jury. And Richard's a tough guy, he's a brawler, and doesn't suffer fools. But he was always very good, just like Donald Judd, he was always very good with landscape architects. He was gentle with them. Whereas at Yale, I've been in juries of his and he's a brute.

So anyway, George had done this fence and he'd painted the fence just like Frank Stella, same colors, didn't change it much. So there were comments made and Serra said, there were some very good things about this, this is nice and so on and so forth, as I say, being very generous. And he said, but you can't just take someone's brand and put it up.

And George said well why not. Anyway, push came to shove and they were going to hit each other. And Frank grabbed Serra and I grabbed George and it was just one of the most interesting juries I've ever had. [Laughter] I always wondered, you know sometime when I'm having a drink with George, I'd ask him if he remembered that jury. He'd probably still be belligerent. But he was so smart.

SM Well, his desk crits were the best desk crits I ever had and, in a way, of that project from his studio it is still the site plan I'm most proud of.

PW Well, he was confident enough and intellectually secure enough to be combative. And he wasn't afraid of the students. An awful lot of faculty are afraid of the students.

SM This makes me think about your own career—I mean you started so young and you practiced for such an extended period of time.

PW One of my favorite artists, an artist who I admire—and it goes beyond admiring the product—is Frank Stella. Because Frank would try something, and then he'd think of something, and he'd try something different. And for a long while he was so popular, I mean the stuff was just like hot cakes, it was selling. And then he got into these three-dimensional off the wall things, series called the birds. And nobody liked them, and they didn't sell. And he started making bigger ones, which were even more expensive and hard to put in an apartment or house, so the museums were the only ones who could afford them. But he kept on and he's still doing that.

He had a big show at the opening of the Whitney in New York last year and we all went. A lot of people I know were there, a lot of designer friends, Martha went. We talked about the show and how we loved the first part, just like the world did, and we didn't like the second part of the show. And unfortunately Frank had said, "I don't care about the old stuff, I want the new stuff up." And so two thirds of the show was the new stuff and a little bitty that was sort of squeezed over in the corner were his masterpieces. Everybody said, "Why didn't they just feature the stuff that was important?" And I had that in my head.

I have a friend in New York who's sort of an art maven and sort of critical and he said, "Well you know, Stella is always more interested in the new, and he's less interested in what he did when he was twenty-five or thirty." And I thought, that's the right way to look at it. Whether he's right or wrong, whether he's selling or not selling, that's the way to look at it. And it was a disappointing show, but it was the show Stella wanted to have. And I probably would feel the same way for myself.

SM Maybe just one final question. Thinking about the forms of practice, it seems to me that talent isn't enough. Clearly you had the talent there from the beginning, and you had the intellectual curiosity, and then in your early career gained the proficiency and expertise. But you also had gone through different phases of growth with SWA to then make a conscious decision about essentially designing the practice that you wanted. In my observation the mid-scale practice can be tremendously challenging to maintain. And yet it would seem to me the highest-quality work, the iconic work is only produced by these kinds of practices.

PW Ok, that's a good frame. There is a kind of practice that used to be called an academic practice. And I would call Rich's practice an academic practice. It's where you're primarily a teacher but you had some time. And you had very little economic threat because, you know you had a salary and you had a place to go to on Monday mornings and so forth. So that's why many painters, many sculptors have to teach. And some of them stop making things because the teaching takes over, and some do more academic things—they write, they photograph—but the academic practice is really just a step above an art studio. Some very good people like Tommy Church never had more than five people in their office.

Then you have what I would call the medium-sized office. Could be ten or fifteen, could be thirty. Once staff gets above thirty it starts becoming something else where there has to be growth. A twenty-five or thirty person kind of office can afford the services they need. Another characterization of this sort of practice is it's structured to have a master designer or two, then some key assistants, and then young people who are getting training. That is a kind of balancing. Through all the time that I have been practicing, even when I worked for Halprin or worked for Sasaki, we never really got big. It wasn't until SWA that I experienced what Hid would call a corporate practice.

So then you have the corporate practice. And the corporate practice is very hard to keep design-oriented. Because you have this need to pump out enough stuff and it leads to branch offices for marketing purposes, not really for regional understanding purposes. George [Hargreaves] is a good example. Gary Hilderbrand has an office and a half, really about thirty people. But when you get to forty you really need an office and a half and maybe you need two branches, and so forth. [James] Corner's finding that out. It's getting bigger, the work's not as interesting. And Corner's not a very good manager. He has learned how to make money, he's been able to get a lot of work, but he hasn't been able to do a lot of good work. So I see the reduction of the value of the work. And these are hard things. I don't speak about them lightly. They're hard things. It's not that a thirty person office doesn't have any economic determinism, it does. And it's a problem because you have to limit the size because you know that if you get up to that next size you're on a growth kick, whether you want it or not.

So those are the three kinds of practices. I started off working for people and working myself in this small studio thing, I wasn't teaching but it was small. It was small because nobody wanted anything, I didn't have any contacts. And then I got up with Sasaki to the mid-size. We were generally around twenty-five or thirty. And I was very comfortable in that. And what was nice about it, because Hid was teaching, he had made a sort of combination, it wasn't so much that the teaching allowed him to have economic security, but it fed students into the system. And Hid wasn't the one who did the things. He was a great manager—was everything that Corner isn't—but he wasn't the idea giver. He was the concept giver but not the idea giver. And he had a whole group of us, really good people. Stu Dawson, Bill Johnson, you know really good people worked there. And still do.

SWA was able to attract GSD people. And some stayed and became partners or some worked for a while and then had their own office like George. Bill Callaway, for instance, stayed. But they were all tied to Harvard.

We have a program in the summer that gives us six or seven students, and we take a few from Harvard, one or two from Penn, University of Virginia, those are the schools that feed us students, feed us employees over time.

One of the things SWA does is hire a tremendous number of Chinese people working as junior staff. And of course [SWA Chairman] John Wong speaks Chinese

and has been key to finding these people. And they have an office in Shanghai.

SM Which is how I got to Asia, because I was with SWA opening that office.

PW What's interesting about it is that for a corporate practice they do very good work. I don't like everything they do, it's not as inventive as it used to be, but I do credit them. I think they are a good office. Generally speaking, there are a number of good offices. Not an infinite number, but there are twenty or thirty offices that I have real respect for and watch what they do. But as they get up to being fifty people I notice that the quality of work, the number of ideas per square foot goes down. I'd had that experience with SWA, and it's the nature of the beast.

Peter Walker and Scott Melbourne, 2019.

PROJECT CHRONOLOGY

Note: The projects referenced in this book are listed in chronological order. Chapter and page numbers indicate where they have been treated. Coordinates shown in parentheses may be used to find project locations with online mapping services, listed in decimal degrees format for ease of use.

1950s

#

Foothill College (37.362, -122.129; Los Altos Hills, California, USA): Ch. 3, Ch. 6   p. 35, 36–38, 41, 122, 150

1960s

#

Fashion Island Center (33.616, -117.875; Newport Beach, California, USA): Ch. 7   p. 56, 57, 78, 79, 90, 91

#

Alcoa Plaza (37.796, -122.399; San Francisco, California, USA): Ch. 4   p. 24, 61–63

#

Sydney Walton Square (37.797, -122.399; San Francisco, California, USA): Ch. 3   p. 24, 38, 39, 122

1970s

#

Weyerhaeuser Headquarters (47.296, -122.299; Federal Way, Washington, USA): Ch. 3   p. 24, 52–55, 156, 157

#

Security Pacific National Bank (34.053, -118.253; Los Angeles, California, USA): Ch. 7   p. 24

#

IBM Santa Teresa (37.196, -121.749; San Jose, California, USA): Ch. 4   p. 58, 59, 70, 71, 150

#

Marlborough Street Roof Garden (Cambridge, Massachusetts, USA): Ch. 2   p. 26, 27

#

Cambridge Center Roof Garden (Cambridge, Massachusetts, USA): Ch. 2   p. 26, 27

1980s

#

Necco Garden (Cambridge, Massachusetts, USA): Ch. 2   p. 27

#

Burnett Park (32.750, -97.333; Fort Worth, Texas, USA): Ch. 3, Ch. 5   p. 44, 45, 94, 149

#

Tanner Fountain (42.376, -71.116; Cambridge, Massachusetts, USA): Ch. 7   p. 144, 145, 146

#

IBM Solana (32.981, -97.174; Westlake and Southlake, Texas, USA): Ch. 5, Ch. 6   p. 86–97, 114, 115, 126, 149

1990s

\#

South Coast Plaza Town Center (33.691, -117.884; Costa Mesa, California, USA): Ch. 3, Ch. 7

\#

IBM Japan Makuhari Building (35.653, 140.043; Makuhari, Chiba Prefecture, Japan): Ch. 5, Ch. 6

\#

Circular Park at Nishi Harima Science Garden City (34.932, 134.442; Nishi Harima, Hyogo Prefecture, Japan): Ch. 3, Ch. 5

\#

Center for Advanced Science and Technology (34.931, 134.444; Nishi Harima, Hyogo Prefecture, Japan): Ch. 5, Ch. 6

\#

Boeing Longacres Industrial Park (47.462, -122.234; Renton, Washington, USA): Ch. 5

\#

Library Walk, University of California at San Diego (32.879, -117.238; San Diego, California, USA): Ch. 3, Ch. 5

\#

Toyota Municipal Museum of Art (35.080, 137.151; Toyota City, Aichi Prefecture, Japan): Ch. 3, Ch. 5, Ch. 6, Ch. 7

\#

Children's Pond and Park (32.710, -117.164; San Diego, California, USA): Ch. 3, Ch. 7

2000s

\#

Novartis Campus (47.573, 7.579; Basel, Switzerland): Ch. 5, Ch. 6

\#

Sony Center (52.510, 13.373; Berlin, Germany): Ch. 5

\#

Saitama Plaza (35.893, 139.631; Saitama, Japan): Ch. 4, Ch. 7

\#

Center for Clinical Science Research, Stanford University (37.432, -122.177; Palo Alto, California, USA): Ch. 6

\#

Medical Center Parking Structure 4, Stanford University (37.434, -122.178; Palo Alto, California, USA): Ch. 4

\#

Nasher Sculpture Center (32.788, -96.800; Dallas, Texas, USA): Ch. 6, Ch. 7

\#

Jamison Square (45.529, -122.682; Portland, Oregon, USA): Ch. 6, Ch. 7

2010s

\#

National September 11 Memorial (40.711, -74.013; New York, New York, USA): Ch. 4, Ch. 7   p. 28

\#

Marina Bay Sands Integrated Resort (1.284, 103.860; Singapore): Ch. 4   p. 72–75

\#

GSB Knight Management Center, Stanford University (37.428, -122.162; Palo Alto, California, USA):
Ch. 3   p. 50, 51

\#

University of Texas at Dallas (32.985, -96.748; Dallas, Texas, USA): Ch. 7   p. 96, 97, 136, 137, 139

\#

Newport Beach Civic Center and Park (33.611, -117.871; Newport Beach, California, USA): Ch. 3, Ch. 5   p. 56, 57, 78, 79, 90, 91

\#

Barangaroo Reserve (-33.857, 151.201; Sydney, Australia): Ch. 4   p. 76, 77, 149

BIBLIOGRAPHY

Alofsin, A. (2002). *The Struggle for Modernism: Architecture, Landscape Architecture and City Planning at Harvard* (1st ed.). New York: W. W. Norton.

Amidon, J. (2006). *Peter Walker and Partners : Nasher Sculpture Center Garden* (Source Books in Landscape Architecture; 3). New York: Princeton Architectural Press.

Birnbaum, C., and Karson, R. (2000). *Pioneers of American Landscape Design*. New York: McGraw Hill.

Kassler, E., and Museum of Modern Art (1984). *Modern Gardens and the Landscape* (rev. ed.). New York: Museum of Modern Art.

Lampugnani, V. M., and Hoiman, S. (eds.) (2009). *Novartis Campus: A Contemporary Work Environment: Premises, Elements, Perspectives*. Stuttgart: Hatje Cantz.

Mozingo, L. (2011). *Pastoral Capitalism: A History of Suburban Corporate Landscapes* (Urban and Industrial Environments). Cambridge, MA: MIT Press.

Petschek, P., and Hochschule für Technik Rapperswil, Institut für Geschichte und Theorie der Landschaftsarchitektur. (2008). *Grading for Landscape Architects and Architects*. Basel: Birkhäuser.

Platt, K., and SWA Group. (1992). *Landscape Design and Planning at the SWA Group*. Tokyo: Process Architecture.

Reed, D., Hilderbrand, G., Kramer, E., and Reed Hilderbrand Associates (2012). *Visible, Invisible: Landscape Works of Reed Hilderbrand*. New York: Metropolis Books.

Simo, M. L. (1999). *100 Years of Landscape Architecture: Some Patterns of a Century*. Washington, D.C.: ASLA Press.

Simo, M. (2001). *The Offices of Hideo Sasaki: A Corporate History*. Berkeley: Spacemaker Press.

Treib, M. (ed.). (1993). *Modern Landscape Architecture: A Critical Review*. Cambridge, MA: MIT Press.

Treib, M. (2015). "Complexity and Spectacle: Three American Landscapes of the 1980s." *Journal of Landscape Architecture*, 10(3), 6–19.

Tunnard, C. (1950). *Gardens in the Modern Landscape* (2nd ed.). London: Architectural Press.

Walker, P. (1997). *Minimalist Gardens*. Berkeley: Spacemaker Press.

Walker, P., and Gillette, J. B. (2005). *Peter Walker and Partners: Defining the Craft*. San Rafael, CA: ORO Editions.

Walker, P., and Reed, H. (1980). *Art Into Landscape: An exhibit organized by Peter Walker in conjunction with Harold Reed;* April 29—May 24, 1980. New York: Harold Reed Gallery.

Walker, P., Hilderbrand, G., Crandell, G., and Gillette, J. (2016). *PWP Landscape Architecture: Building Ideas.* San Rafael, CA: ORO Editions

Walker, P., Jewell, L., and Harvard University Graduate School of Design. (1990). *Peter Walker: Experiments in Gesture, Seriality, and Flatness.* New York: Rizzoli.

Walker, P., and Simo, M. L. (1996). *Invisible Gardens: The Search for Modernism in the American Landscape.* Cambridge, MA: MIT Press.

Way, T. (2015). *The Landscape Architecture of Richard Haag: From Modern Space to Urban Ecological Design.* Seattle: University of Washington Press.

Zimmermann, A. (2011). *Constructing Landscape: Materials, Techniques, Structural Components* (2nd, rev. and expanded ed.). Basel: Birkhäuser.

ABOUT THE AUTHORS

SCOTT JENNINGS MELBOURNE is an Assistant Professor of Landscape Architecture at the University of Hong Kong whose research advances design scholarship. His studio teaching initiatives have explored the potential for design to inform decision making in developing regions, with a particular focus on Southeast Asia. Melbourne possesses more than a decade of experience in professional practice, including roles at Richard Haag Associates, Mithun Architects, and SWA Group. Melbourne holds the degree of Master in Landscape Architecture with distinction from Harvard University and Bachelor of Landscape Architecture from the University of Washington. While at Harvard he was a Penny White recipient and awarded the Norman T. Newton Prize.

NIALL KIRKWOOD DSC FASLA is Professor of Landscape Architecture and Technology, Co-Director of the GSD Technology Platform and Associate Dean for Academic affairs at the Harvard Design School, Cambridge, Massachusetts, USA.

ACKNOWLEDGMENTS

There are many to thank.

Dorothy Tang for suggesting I "write that up," Niall Kirkwood who encouraged I expand these notions into a full manuscript, and editor Ria Stein who immediately recognized the subject's value and expertly guided this book to its final form.

My mentors, especially John Stilgoe and David Streatfield, who equipped me with ways of seeing that made this exploration possible.

Peter Walker himself, who was generous in his availability for this independently developed study.

Talented assistants made key contributions along the journey: Emma Mendel, Ellena Wong, Haylie Shum, and Wiley Ng.

Jessica Bridger and Brice Maryman provided insightful feedback on early chapters.

The support of colleagues and my university proved essential, while the steadfast encouragement of my family and especially Jenny helped me persevere.

Finally, place matters: I am grateful for the village of Peng Chau within which this text was written, and the great city of Hong Kong that endured significant challenges as the project was completed.

ILLUSTRATION CREDITS

GRAPHIC DESIGN, LAYOUT AND TYPESETTING: Vera Pechel
COPY EDITING AND PROJECT MANAGEMENT: Ria Stein
PRODUCTION: Heike Strempel

COVER: IBM Japan Makuhari Building
COVER PHOTOGRAPH: Scott Melbourne

PAPER: 135 g/m² Condat Matt Perigord
LITHOGRAPHY: bildpunkt Druckvorstufen GmbH
PRINTING: optimal media GmbH

Library of Congress Control Number: 2020930230
Bibliographic information published by the German National Library
The German National Library lists this publication in the Deutsche Nationalbibliografie;
detailed bibliographic data are available on the Internet at http://dnb.dnb.de.

ISBN 978-3-0356-1610-1
e-ISBN (PDF) 978-3-0356-1583-8

© 2020 Birkhäuser Verlag GmbH, Basel
P.O. Box 44, 4009 Basel, Switzerland
Part of Walter de Gruyter GmbH, Berlin/Boston

Printed on acid-free paper produced from chlorine-free pulp. TCF ∞

Printed in Germany

9 8 7 6 5 4 3 2 1   www.birkhauser.com